Social and Cultural Aspects of Circular Economy

This collection of essays brings together discussions arguing that the circular economy must be linked to society and culture in order to create a viable concept for remodelling the economy. Covering a diverse range of topics and regions, including cities and living, food and human waste, packaging and law, fashion, design, and art, this book provides a multi-layered examination of circularity.

Transitioning to a circular economy, reducing resource input and waste, and narrowing material and energy loops are becoming increasingly important targets to combat decades of unsustainable models of consumption. However, they will require a significant shift in social and cultural thinking and these dimensions have not yet been factored into policy debates and frameworks. While recognising the key role of individual consumers and their behaviours, the book goes beyond this singular perspective to provide equal focus on institutional and political structures as necessary drivers for real change.

Social and Cultural Aspects of the Circular Economy argues for a social and solidarity economy (SSE) to combine individual actions with a wider cultural shift. It will be an important read for scholars, researchers, students, and policymakers in the circular economy, waste studies, consumption, and other environmentally focused social sciences.

Viktor Pál is a Researcher at the Department of Cultures at the University of Helsinki, Finland, and has the title of Docent at the University of Tampere, Finland. He also serves as Coordinator at the Helsinki Environmental Humanities Hub.

Social and Cultural Aspects of the Circular Economy

Toward Solidarity and Inclusivity

Edited by Viktor Pál

Routledge
Taylor & Francis Group
LONDON AND NEW YORK

earthscan
from Routledge

First published 2022
by Routledge
4 Park Square, Milton Park, Abingdon, Oxon OX14 4RN

and by Routledge
605 Third Avenue, New York, NY 10158

Routledge is an imprint of the Taylor & Francis Group, an informa business

British Library Cataloguing-in-Publication Data
A catalogue record for this book is available from the British Library

Library of Congress Cataloging-in-Publication Data
A catalog record has been requested for this book

ISBN: 978-1-032-18580-4 (hbk)
ISBN: 978-1-032-18583-5 (pbk)
ISBN: 978-1-003-25524-6 (ebk)

DOI: 10.4324/9781003255246

Typeset in Times New Roman
by Deanta Global Publishing Services, Chennai, India

Contents

Figures

Tables

Contributors

Maria Åkerman acts as a Principal Scientist at VTT Technical Research Centre where her focus area is the sustainability of socio-technical changes. She has a multidisciplinary social scientific background in environmental policy, science and technology studies, and economics and her key research interests include socio-technical transitions, sustainability governance, and politics of environmental knowledge production.

Sanne Bor is a postdoctoral researcher and project manager at LUT University, Finland, involved in a large multidisciplinary research project called Package Heroes (2019–2023). Her current research focuses on topics related to collaboration among organisations, including meta-organisation, self-steering, horizontal organising, and collective decision-making in the context of sustainability transitions and environmental conflict.

Iris Borowy is a Distinguished Professor at Shanghai University, China. She is also a founding director of the Center for the History of Global Development at that university. Her publications include *Defining Sustainable Development for Our Common Future: A History of the World Commission on Environment and Development (Brundtland Commission)*, published in 2014.

Sultan Çetin is an architect and engineer with a construction and design management background. She is currently a PhD researcher at the Delft University of Technology in the Netherlands. Her research looks at digital technologies and their application in the circular built environment. She is the co-founder of DiCE Lab (Digital Circular Economy Lab).

Ari Jokinen is a senior research fellow in Environmental Policy in the Faculty of Management and Business at Tampere University, Finland. He has studied sustainability policy focusing on urban issues, natural resources, biodiversity, circular economy, and the politics of nature. He is particularly interested in forms of organising between humans, technologies, and non-humans in sustainability processes.

Pekka Jokinen is a Professor of Environmental Policy at Tampere University, Finland. His research focuses on environmental governance, social change, and sustainabilities. His current work concerns climate policy and urban circular economy. Jokinen has written on a wide range of issues in environmental politics and policy.

Katariina Koistinen holds a PhD (2019) in Environmental Management from LUT University, Finland. Currently she is working as a postdoctoral researcher at the University of Turku, Finland. Her research focuses on actors in facilitating sustainability change. Her research interests include sustainability transitions, sustainable management, and theories of agency.

Sabine Lettmann is a Senior Lecturer at the Institute of Jewellery, Fashion, and Textiles at Birmingham City University, UK, with international teaching experience in fashion concept development, circular design systems, and costume design. Since 2001, she works as a freelance fashion designer and consultant which influences her research interests around the subject of circular fashion and design education.

Marileena Mäkelä is a Senior Lecturer at the Jyväskylä University School of Business and Economics (JSBE) and the School of Resource Wisdom at the University of Jyväskylä, Finland. She teaches corporate sustainability at JSBE and the Turku School of Economics. Her research interests include sustainability reporting, circular economy, and futures images.

Carlos Miret is a Research Support Assistant at the Sussex Sustainability Research Programme (SSRP), an interdisciplinary programme based on a partnership between the University of Sussex and the Institute of Development Studies, UK. Carlos holds an MSc in Food Sustainability and Natural Resources Management.

David Ness is an Adjunct Professor within UniSA STEM, University of South Australia, focusing upon rebalancing resource consumption in the built environments of the Global North and South. He served on the Committee for the World Resources Forum 2021, and was formerly an Advisor to UNESCAP and UN-Habitat on eco-efficient, equitable, and inclusive infrastructure.

Erkki-Jussi Nylén is a PhD researcher in environmental policy at Tampere University, Finland. His research has focused on policy processes, the role of ideas and concepts in policymaking, and sustainability transitions. Currently Nylén is finishing his doctoral dissertation in which he analyses the circular economy as a policy concept, and how it is actualised to practices through policy processes.

Greg O'Shea is a researcher at LUT University, Finland, and Executive-in-Residence at Aalto University School of Business, Finland. He is currently researching the business model and entrepreneurial ecosystem logic within the Package Heroes project. His current research is around sustainability-focused and regional entrepreneurial ecosystems, and sustainability transitions.

Tiina Onkila works as an Associate Professor at the University of Jyväskylä, Finland. Her education background is a doctoral degree since 2009 in corporate environmental management and she also holds an adjunct professor position at the Turku School of Economics, Finland. Her research interests relate to meanings of sustainability in business, in particular stakeholder relations, employee-CSR relations, sustainability agency, and circular economy.

Viktor Pál is a Researcher at the Department of Cultures at the University of Helsinki, Finland, and holds the title of Docent at the University of Tampere, Finland. He also serves as a Coordinator at the Helsinki Environmental Humanities Hub. His first book *Technology and the Environment in State-Socialist Hungary: An Economic History* was published in 2017.

Taina Pihlajarinne is a professor at the Helsinki University Faculty of Law, Finland. Her research interests are focused on the relationship between sustainability and private law, especially intellectual property rights. At this moment she leads the research project "Shaping, fixing and making markets via IPR: Regulating sustainable innovation ecosystems" funded by the Academy of Finland.

Raysa França is the Regional Director for Europe and Central Asia at Youth4Nature and a co-founder at Symbiosis Tampere. She is a storyteller and an advocate for diversity and inclusion in the climate movement. Based in Finland, and from Brazil, Raysa is especially curious about links between cultural diversity and biodiversity. With more than six years of professional and volunteering experience, she has worked in business environments, academia and NGOs. Raysa holds a bachelor's degree in Social Sciences from the Federal University of Minas Gerais, and a master's degree in Administrative Sciences from Tampere University.

Jesus Rosales Carreon is an Assistant Professor at the Copernicus Institute of Sustainable Development, Utrecht University, the Netherlands. He also serves as Programme Leader of the Energy Science master's programme at this institution. His research interests centre around the possibilities to assess the impacts of circular strategies.

Halima Sacranie is the Research Lead of the Housing and Communities Research Group at the University of Birmingham, where she completed her PhD in Urban and Regional Studies in 2012. Her research interests include affordable and sustainable housing models, community investment and social impact. Halima is also Chair of the Housing and Communities Leadership Board at the think-tank Centre for the New Midlands in the UK.

Milla Sarja is a doctoral researcher in the field of Corporate Environmental Management at the Jyväskylä University School of Business and Economics, Finland. Her research interests include circular economy, especially in business context, sustainability transitions, and sustainability agency.

Pia Schmoeckel, with a background in psychology, is interested in topics surrounding behavioural change and public policy, motivated to increase health-enhancing and socially benefitting behaviour. Her work experience at the Policy and Strategy Unit of the Western Cape Government in South Africa has influenced her aim to address challenges with human-centred and social approaches. Following her undergraduate studies, she will pursue a MSc in Behavioural Science at the London School of Economics.

Kirsi Sonck-Rautio is a postdoctoral researcher at the Åbo Akademi University, Finland, working within the multidisciplinary consortium Package Heroes (2019–2023). She has a Doctorate in European Ethnology and she has specialised in consumer cultures and environmental ethnology, especially on global environmental problems such as biodiversity loss, climate change, and plastic waste.

Claudia Stuckrath is a consultant and a researcher at the Copernicus Institute of Sustainable Development, specialising in circular economy. She also works as a specialist for UULabs, Utrecht University Living Labs for Sustainable Development. She has successfully managed innovative engineering and academic projects and has been published in recognised peer-reviewed journals.

Lis Suarez-Visbal is an Ashoka fellow, a consultant and a Doctoral Researcher at the Copernicus Institute of Sustainable Development, Utrecht University, the Netherlands. She specialises in the intersection of circular economy and social impact with a gender lens. She has over 15 years of experience leading initiatives related to circular economy uptake in various countries within the apparel value chain, among other industries.

Satu Teerikangas is Professor of Management and Organisation at the University of Turku (UTU), Finland, and Honorary Professor of Management at the University College London, UK. Her research centres on strategic change and change agents. Her research on mergers and acquisitions is recognised (BBC, *Forbes*). She is editor of the *Oxford Handbook of M&As* (2012) and the *Edward Elgar Research Handbook on Sustainability Agency* (2021).

Nina Tynkkynen is Professor (tenure track) in Environmental Policy and Governance at Åbo Akademi University, Finland. Her research focuses on multilevel governance settings, the politics of knowledge, and the construction of environmental problems and related policies.

Luciana Valio is an artist, researcher, and Professor of Tridimensionality at the Institute of Letters and Arts – ILA, Federal University of Rio Grande (FURG), Brazil. She also serves as the Assistant Coordinator of the Visual Arts Courses at ILA/FURG. She was a postdoctoral researcher in visual arts at the State University of Campinas (UNICAMP), Brazil. Her research field in contemporary art is interested in concepts such as decoloniality, perspectivism, and epistemologies of the south.

Mira Valkjärvi is a doctoral student in the field of management and organisation at the University of Turku, Finland. Her research interests include circular economy, service-dominant logic, the role of motivation and culture in sustainability, and the use of narratives in CE research.

Vilja Varho is a Senior Scientist at the Natural Resources Institute Finland (Luke). She has a doctorate in agriculture and forestry, and a title of Docent in futures research. Varho studies sustainability transitions, particularly those relating to renewable energy and food, through futures research methods and stakeholder co-creation.

Acknowledgements

The current volume is the first publication of the Helsinki Environmental Humanities Hub, a meeting place of scientists and artists at the University of Helsinki that studies the multifaceted cultural and relational dimensions of anthropogenic environmental change.

Contributions in this volume are based on papers presented at the Social and Cultural Aspects of Circular Economy Online International Conference organised by Viktor Pál and held at the Helsinki Environmental Humanities Hub, University of Helsinki, 19–20 November 2020.

The Social and Cultural Aspects of Circular Economy conference and the editor's work on the current volume was generously supported by the Helsinki University Humanities Program led by Mikko Saikku and the Helsinki Institute of Sustainability Science at the University of Helsinki.

1 Social and cultural aspects of circular economy

Towards solidarity and inclusivity

Viktor Pál

The concept of circular economy

The take-make-waste extractive industrial model has created an open system globally which is characterised by the continual influx of new products designed to be used briefly and then discarded. That open system has decisively contributed to the current global environmental crisis and climate change. It is the consensus of the environmentally focused scientific community that the shift to a circular economy (CE) in which resource input and waste, emissions, and energy leakage are minimised by slowing, closing, and narrowing material and energy loops as necessary (Azevedo et al. 2017; Crocker et al. 2018; Gallaud et al. 2016; Jiansu 2018; Matthews and Tan 2011).

Despite the popularity of the CE model, there is little consensus on the definition of the circular economy. Competing definitions have described CE from various perspectives based on extensive research. For example, according to Geissdoerfer et al., a circular economy is "A regenerative system in which resource input and waste, emission, and energy leakage are minimized by slowing, closing, and narrowing material and energy loops. This can be achieved through long-lasting design, maintenance, repair, reuse, remanufacturing, refurbishing, and recycling" (2017: 766.) The definition by Geissdoerfer et al. holds many features in common with other descriptions, such as the one by Ghisellini et al., who focus on the maximum reuse/recycling of materials, goods, and components and the innovation of the entire chain of production, consumption, distribution, and recovery in their definition (Ghisellini et al. 2016). In fact, ideas of the CE often concentrate on the industrial model of getting waste and pollution out of the system and keeping products and materials in use. Thus, the CE concept has enticed attention from various stakeholders because of its apparent economic, technological, financial, and ecological benefits. As a result, CE has been adopted as a national policy by China in 2009 as well as the European Commission in 2015.

Simultaneously, CE has become a magnet for a growing number of scholars who joined the discourse over the shift to a circular economy. The intensification of the CE discourse has led to the identification of significant research gaps. It has been acknowledged by a handful of scientists that CE is not a mere

DOI: 10.4324/9781003255246-1

economic-technological system without human implications, but has social and cultural dimensions as well (Fitch-Roy et al. 2020; Hobson 2020; Lazarevic and Valve 2017; Moreau et al. 2017; Murray et al. 2017; Schöggl et al. 2020). In this vein, scholars have often attributed the relatively slow adoption of CE to specific social and cultural reactions. For example, Walter R. Stahel, one of the founders of European circularity, assumed that the lack of familiarity and fear of the unknown was behind the societal passivity and even hostility against CE (Stahel 2016). The lack of consumer interest in CE has been often recognised to be intertwined with a lack of support networks for a wide range of actors to implement a CE strategy and innovations (Kirchherr et al. 2017). As well, the lack of openness from actors has been often attributed to the lack of integration of particular social and cultural aspects, e.g., housing, food systems, working conditions and workers' rights, and international law within the CE discourse (Geisendorf and Pietrulla 2018). To sum it up, on the one hand, the CE concept struggles to gain wide social and cultural acceptance in many societies and sectors around the globe, and it is threatened to remain an econo-technological fix that does not target the core issues of the global environmental crisis while, on the other, the bulk of research is still focusing on the econo-technological layers of CE leaving large gaps in the core of the CE – namely the social and cultural layers – unexplored.

Circular economy as a socio-cultural phenomenon

Authors of this volume recognise this controversy and acknowledge that to make CE the dominant system for human societies is not primarily a question of econo-techno fixes. Thus, the essays in this volume suggest that linear economic systems are rooted in human action, which in return stems from the values, ethics, and attitudes of human societies and cultures (Pál 2017; Pál 2021a; Pál 2021b). The importance of the human layers in CE is grave and without the social and cultural layers incorporated, the CE discourse will not be able to generate substantial change towards sustainable human systems (Korhonen et al. 2018).

This situation has already led some experts to call for understanding CE not as an econo-technological system detached from our societies, but as an integral part of both human culture and the ecosystem via a value-based approach (Mies and Gold 2021). In this vein, scholars have already identified consumption as being the dominant mode of human action that impacts CE. Research that has been carried out in this respect has either aimed to identify factors that drive human consumption to a CE circle, or to discuss the integration of consumer perspectives into the design process (Camacho-Otero et al. 2018; Schröder et al. 2019). This volume takes a third approach as its point of departure, namely that the complexities of human actions, values, beliefs, and ethics need to be better understood and interpreted for a successful shift to CE. The essays presented here acknowledge the key role of consumption in CE, but also point out the richness and complexity of human actions that make an impact on CE and the global ecosystem in general. In addition, essays of this volume identify the disparities and wealth gaps – both locally and globally between the Global North and South– as key solidarity and inclusivity relations points to a successful CE turn.

It is still unclear how and especially who will benefit from the new opportunities offered by CE, and the scientific community still understands little of how CE will reduce global disparity and contribute to the creation of a society characterised by more solidarity and inclusion both on local and global scales. Thus, the authors of this volume wish to contribute to this scientific urgency and map out the diverse ways CE can promote solidarity and inclusivity across our societies, while contributing simultaneously to environmental improvements globally.

To achieve this goal, chapters in this volume explore key aspects of human life and the opportunities represented in new culturally and socially tuned CE systems. As a result, authors of this volume introduce many aspects of the triple challenge of reducing environmental degradation, creating new opportunities for humans while reducing inequalities, and meanwhile providing a multilayered analysis and complex scientific suggestions for achieving these goals.

In the opening essay Marileena Mäkelä, Tiina Onkila, Satu Teerikangas, Milla Sarja, Mira Valkjärvi, and Katariina Koistinen situate the cultural aspects of the CE implementation based on qualitative interview data by addressing what is the role of cultural factors in the CE transition. This paper offers a cultural perspective to CE transitions, especially those of human interactions in a society. The authors of this chapter consider the concept of culture as a key factor within the shift to CE; however, they acknowledge that culture is a multidimensional concept and takes different meanings in different contexts. Based on a wide array of interviews this chapter suggests that first, a change in values is needed to reach a CE. However, that change in attitudes seems to be emerging slowly and so awareness and raising awareness about CE is a key feature among stakeholders, who need new and more information to increase their ability for the cultural change to CE. Finally, the authors of this chapter identify the individual decision maker's role, and that cooperation and solidarity are key success factors when it comes to the shift to CE.

In Chapter 3, Raysa França, Erkki-Jussi Nylén, Ari Jokinen, and Pekka Jokinen investigate the urban policy context of social and solidarity economy (SSE) which in their view can and should be implemented within the shift to CE in an urban context. This chapter argues that the institutional and policy environment of cities are essential to CE transition, as over 50% of the human population live in urban areas and 60–80% of natural resources are consumed in cities. This chapter analyses the notions of SSE with the goal to understand how it can inform a socially just and environmentally sound urban transition to CE. The study is based on urban policy analyses in Latin America and Europe.

The next chapter by Halima Sacranie and Sultan Çetin also investigates the interrelations of SSE and urban CE policies and focuses on circular social housing organisations (SHOs) and their social impact fostering SSE with a focus on disadvantaged communities, while encouraging households to embrace sustainable modes of dwelling. This study argues that when individual actions are intertwined with circular social housing policies and institutions, they can promote social inclusion, tenant engagement, and empowerment, which in return foster SSE and contribute to the shift to CE. This chapter is based on longitudinal case studies of circular social housing projects in the Netherlands, France, Belgium, and the UK.

Chapter 5 by Carlos Miret is the third chapter in this volume that focuses on the urban context of CE and is in dialogue with Chapters 3 and 4 by considering the circular city transition via a CE urban food transformation. This essay takes the UK municipalities of Hove and Brighton as case studies to develop a city-region circular food road map, in which local food initiatives are mapped in preparation for a CE food system transformation. The analysis in this chapter is suggesting that CE and SSE should be intertwined with the localisation of resources as a potentially successful strategy to build community food resilience. Both individual and policy as well as institutional actions are considered when planning surplus food redistribution as a strategy to reduce food waste and food poverty. By doing so, this chapter highlights a caveat: whilst the CE holds potential as a new frame for food system change and cross-sectoral governance, a narrow focus on food waste reduction targets risks neglecting the need for structural changes.

Intertwined with the dimensions of food in CE, Chapter 6 by Iris Borowy is considering the institutional and policy contexts of human waste, a matter that is predominantly produced in cities globally today. This study argues that many historical examples may be useful when considering the circulation of human waste degrades. This chapter considers a handful of cultural aspects of defecation, many of which stigmatise human faecal matter, as well as analyses the policy and institutional aspects of the collection, transportation, and disposal of human waste on a global scale. This paper maps out which social and cultural factors stand in the way of faecal circular systems and how international institutions such as the World Health Organization could improve the efficacy of their actions when it comes to the CE of human waste. Lastly, this chapter suggests that health risks are often argued to be the problem when it comes to the circularity of human waste. In fact, the real issue is the deep-seated institutional and policy considerations which – intertwined with individual human actions – create a context where artificial fertilisers are cheap and dealing with human waste is stigmatised.

In Chapter 7, Kirsi Sonck-Rautio, Sanne Bor, Greg O'Shea, Nina Tynkkynen, Vilja Varho, and Maria Åkerman connect human food systems to human waste (Chapters 5 and 6) and consider the institutional, policy, and consumer aspects of food packaging, which produces a major proportion of the global waste, plastic waste in particular. However, food packaging is a double-edged tool as it is needed in order to promote ecologically sustainable development, as it is an efficient way to prevent food waste. This chapter argues that we need to reduce both food and packaging waste via creating more sustainable packaging solutions, the challenge of which is not primarily a technological one. The authors of this chapter suggest that because of the underlying layers of institutions, policies, and consumer behaviour, the roles of different actors in the transition process towards CE food packaging should first be considered, instead of the technological questions put forwards. Based on a wide array of interviews with stakeholders in Finland, this chapter analyses how consumers, policy-makers, NGOs, and businesses view their possibilities and barriers, power and responsibilities to promote the transition towards CE food packaging.

In Chapter 8, Taina Pihlajarinne connects to packaging issues and examines the legal framework of product lifespans from the perspective of private law where consumer law and intellectual property law serve as examples. This chapter analyses the notion of "expected" and "normal" lifespan of products and their relation to CE as well as the role of consumers, legislators and courts. Pihlajarinne argues that to promote a CE, the contents and limits of product lifespan should be redefined to reflect the call for sustainability. Although Nordic countries' consumer attitudes are more favourable, Pihlajarinne asserts that the legal concepts of lifespan should be further developed by legislators and courts to create new incentives for products with a more sustainable lifespan. Thus, this chapter analyses the complex relationship between legal-, social-, and moral norms and proposes a new, SSE-CE-driven framework especially from EU-, European-, and Nordic law perspectives.

In the following chapter Lis Suarez-Visbal, Claudia Stuckrath and Jesús Rosales Carreón connect with both the various aspects of consumption such as food packaging (Chapter 7) and the legal grounding of CE (Chapter 8) and focus on the CE aspects of the apparel industry, which is often stigmatised by the take-make-waste model, unfair working conditions, and environmental degradation. At the same time, they argue, purchases of apparel consumers globally are essential for ensuring the livelihood of millions of workers in developed and developing countries. In this contradictory institutional environment, a growing number of apparel businesses are adopting CE as their framework of operations to achieve sustainability. However, currently, there is a lack of framework to assess the social impacts generated by the adoption of circular practices by apparel businesses. Thus, the authors of this chapter propose a new institutional framework based on the quality of jobs, community livelihood, and gender equality and inclusion in the apparel sectors in which both businesses' actions and policies can be guided by SSE and CE principles. They test this framework with the Dutch apparel value chain.

In Chapter 10, Sabine Lettmann and Pia Schmoeckel consider the sustainable fashion sector business approaches via five entrepreneurial projects from the Netherlands, Germany, Denmark, and Italy. As the process of design is not yet specified in the definition of a CE per se, this paper builds on design activism and social design to investigate ways of fostering a shift to sustainable economic systems intertwined with an active formation of multicultural and inclusive communities. Thus it emphasises the active formation of multicultural and inclusive design communities, as well as the role of refugees, disabled people, and migrants, whilst focusing on craftsmanship as an underpinned process of social change. As a result, this study seeks to reflect on business structure, product development, and manufacturing which are supporting SSE and CE. The authors of this chapter also identify a collective mindset embracing equality as a form of social inclusion beyond the technical and business aspects of CE.

The next chapter builds on Chapter 9 on fashion and Chapter 10 on design and analyses the intersections and potential influence of contemporary art as well the traditional knowledge of indigenous peoples on civilian construction in Brazil.

According to Luciana Benetti Marques Valio, contemporary art often reflects on the impacts of the take-make-use-dispose-pollute system intertwined with the construction business, which is in sharp contrast with the nature-integrated construction system used by the indigenous peoples in Brazil. Therefore, when contemporary artists are set to analyse the construction works of indigenous communities, they are also considering alternative ways of life and alternative ways to consider the environment than of the mainstream Western view. Hence, this chapter will consider the cosmovisions of indigenous communities and the integrated way of indigenous construction and its representation in contemporary Brazilian art and, by doing so, will reflect on how contemporary art and artists may facilitate SSE and CE discourses.

The last chapter of the volume, by David Ness, considers SSE-CE from a critical perspective and calls for more solidarity on the global scale, which the author sees as key in going beyond the current economic model. Hence, this chapter holds unbalanced resource consumption at a global level as the key issue at stake, which it considers as an issue of global ethics. This chapter argues that the Global North should not only aim for CE but go beyond CE and reduce its consumption by as much as 90% to enable the fast-growing South to double its resource use as it strives to emerge from poverty and address systematic inequalities. This global solidarity approach is described in this chapter within the context of the concept "shrink and share", which is based on CE-SSE principles but tackles the painful global inequalities maintained by individual consumers and institutional-policy frameworks. The chapter takes the global construction industry as its case study and central to this ambition is the requirement for economic growth and built assets to be seen as supporting social and human services, with recognition of the concepts of "ecological ceiling" and "social foundation".

Although contributions in this volume cover diverse global regions and introduce a wide range of social and cultural aspects of CE, including cities and living, food and human waste, packaging and law, fashion, design, and art, they recognise the key role of individual actions in CE and aim to interpret the shift to CE as being closely connected to the actions of human agents. Having said that, chapters in this volume are aware of the gap in previous research which has mainly seen human agents either as individuals or as subjects to institutional and policy frameworks around them. In addition to the individual and the wider institutional and policy environments, essays in this volume considered aspects of emotional intelligence in the forms of solidarity and inclusivity as aimed at future CE societies. Thus, this volume suggests that simultaneously by focusing on the agency of individual humans and social structures, equal attention should be targeted towards cultural factors. Thus, the social and solidarity economy (SSE) will have a decisive role in negotiating that dual focus on both individual actions and the wider societal and cultural environment in which human actions are embedded together. Thus, it is suggested that prioritising solidarity and inclusivity both on local and global scales as key values is necessary for the creation of effective circular economy systems in the future.

References

Azevedo, Garrido, Marias, Susana, and Carlos O. João. *Corporate Sustainability: The New Pillar of the Circular Economy* (Hauppauge, NY: Nova Science Publishers, Inc., 2017).

Camacho-Otero, Juana, Boks, Casper, and Pettersen, Ida "Consumption in the circular economy: A literature review." *Sustainability* 10 (2018): 2758.

Crocker, Robert et al. *Unmaking Waste in Production and Consumption: Towards the Circular Economy* (Bingley: Emerald Publishing, 2018).

Fitch-Roy, Oscar, Benson, David, and Monciardini, David. "Going around in circles? Conceptual recycling, patching and policy layering in the EU circular economy package." *Environmental Politics* 29, no. 6 (2020): 983–1003.

Gallaud, Delphine, and Laperche, Blandine. *Circular Economy, Industrial Ecology and Short Supply Chain* (London:Wiley, 2016).

Geisendorf, Sylvie, and Pietrulla, Felicitas. "The circular economy and circular economic concepts: A literature analysis and redefinition." *Thunderbird International Business Review* 60 (2018): 771–782.

Geissdoerfer, Martin, Savaget, Paulo, Bocken, Nancy M.P., and Hultink, Erik Jan "The circular economy: A new sustainability paradigm?' *Journal of Cleaner Production* 143, (2017): 757–768.

Ghisellini, Patrizia, Cialani, Catia, and Ulgiati, Sergio. "A review on circular economy: The expected transition to a balanced interplay of environmental and economic systems." *Journal of Cleaner Production* 114 (2016): 11–32.

Hobson, Kersty. "'Small stories of closing loops': Social circularity and the everyday circular economy." *Climatic Change* 163, no. 1 (2020): 99–116.

Jiansu, Mao. *Circular Economy and Sustainable Development Enterprises* (Singapore: Springer, 2018).

Kirchherr, Julian, Reike, Denise, and Hekkert, Marko. "Conceptualizing the circular economy: An analysis of 114 definitions." *Resources, Conservation and Recycling* 127 (2017): 221–232.

Korhonen, Jouni, Nuur, Cali, Feldmann, Andreas, and Birkie, Seyoum Eshetu. "Circular economy as an essentially contested concept." *Journal of Cleaner Production* 175 (2018): 544–552.

Lazarevic, David, and Valve, Helena. "Narrating expectations for the circular economy: Towards a common and contested European transition." *Energy Research & Social Science* 31 (2017): 60–69.

Matthews, John A., and Tan, Hao "Progress toward a circular economy in China: The drivers (and inhibitors) of eco-industrial initiative Mathews." *Journal of Industrial Ecology*, 15 (2011): 435–457.

Mies, Annika, and Gold, Stefan. "Mapping the social dimension of the circular economy." *Journal of Cleaner Production*, 321 (2021) Online.

Moreau, Vincent, Sahakian, Marlyne, Van Griethuysen, Pascal, and Vuille, François. "Coming full circle: Why social and institutional dimensions matter for the circular economy." *Journal of Industrial Ecology* 21, no. 3 (2017): 497–506.

Murray, A., Skene, K., and Haynes, K. The circular economy: An interdisciplinary exploration of the concept and application in a global context. *Journal of Business Ethics* 140 (2017): 369–380.

Pál, Viktor "Toward Socialist Environmentalism? Scientists and Environmental Change in Modern Hungary" *Environment and History* (2021a) (Fast Track, Online 2021).

Pál, Viktor "Waste and Recycling in Cold War Eastern Europe: The Case of the "Bee" Waste Collection Trust in Hungary" *Discard Studies* (2021b) 12.04.2021.

Pál, Viktor. Technology and the Environment in State-Socialist Hungary: An Economic History. London-New York: Palgrave Macmillan, (2017).

Schöggl, Josef-Peter, Stumpf, Lukas, and Baumgartner, Rupert J. "The narrative of sustainability and circular economy: A longitudinal review of two decades of research." *Resources, Conservation and Recycling* 163 (2020): 105073.

Schröder, Patrick et al. eds. *The Circular Economy and the Global South: Sustainable Lifestyles and Green Industrial Development* (London: Routledge, 2019).

Stahel, Walter R. "The circular economy." *Nature* 531 (2016): 435.

2 Transitioning to the circular economy

Shifting from a technical to a cultural perspective

Marileena Mäkelä, Tiina Onkila, Satu Teerikangas, Milla Sarja, Mira Valkjärvi, and Katariina Koistinen

Introduction

The circular economy (CE) aims to create a system that enables the circulation of resources in society without the creation of waste. As the MacArthur Foundation (2021) defines it, the CE "is based on three principles: designing out waste and pollution, keeping products and materials in use, and regenerating natural systems".

The previous literature on the CE tends to focus only on the creation of economic prosperity while simultaneously considering environmental quality. This was a key finding from the review of CE definitions by Kirchherr, Reike, and Hekkert (2017). They further note that although a shift from the current linear economy to the CE would require a systemic change, the role of social and cultural aspects in this transition is seldom discussed in definitions of the CE. Geissdoerfer et al.'s (2017) review focuses on the comparison between the CE and sustainability, and they notice a lack of attention paid to social aspects in the literature (cf. Murray, Skene, and Haynes 2017). Furthermore, the CE review by Sarja et al. (2021) highlights the need for more studies on human action in the CE.

In order to start addressing this gap, this chapter offers a cultural perspective on CE transitions. The cultural dimension refers to human interactions in a society (Bidney 1944). We argue that the concept of culture is a key factor within CE transitions. Like the CE model itself, culture is a multidimensional concept that takes on different meanings depending on the context. Traditionally, culture has been defined as consisting of multiple elements related to human behaviour in groups, organisations, and societies. Thus, in the traditional definition proposed by Tylor (according to Peterson 1979), culture consists of the knowledge, beliefs, art, morals, law, customs, and habits expected from the members of a society. In this chapter, we use the cultural perspective to address individuals and their interactions in the Finnish societal context.

The research question we address in this chapter is: what is the meaning of cultural factors in the CE transitions? We base our chapter on a large sample of qualitative interview data. In our findings, we focus on the cultural catalysts for and obstacles to CE implementation. Our chapter contributes to the narrow field

DOI: 10.4324/9781003255246-2

studying the cultural CE by illustrating the vital role of this dimension of CE transitions.

In the next section, we cover the theoretical background of our work, namely the CE framework and its cultural dimension. In the methods section, we present our research approach: the interviews and their analysis. In the results section, we consider the cultural aspects emerging from our data. The main theme of our results is the need for change, and this need is discussed from three perspectives: values and attitudes, awareness and knowledge about the CE, and cooperation and solidarity in the CE. We end our chapter with a discussion of our results and offer ideas for future research.

Theoretical background

The concept of the circular economy

The CE concept was introduced in order to address the environmental problems caused by the current linear economy. The linear economy can be described as a "take–make–dispose" system (Gregson et al. 2015; Ellen MacArthur Foundation 2013a), wherein raw materials are converted into final products and, in the end, disposed as waste (Elia, Gnoni, and Tornese 2017; Sauvé, Bernard, and Sloan 2016). The system is based on the existence of "large quantities of easily accessible resources and energy" (Ellen MacArthur Foundation 2013b: 26). As Sauvé, Bernard, and Sloan (2016: 53) describe it, "the circular economy aims to decouple prosperity from resource consumption". Although the CE model tackles global environmental problems, according to the Ellen MacArthur Foundation (2015), it also maintains opportunities for economic growth and job creation.

The CE has recently attracted increasing research attention. Thus, the literature offers multiple definitions for the term (e.g., Schöggl, Stumpf, and Baumgartner 2020). To illustrate, some of these definitions are gathered in Table 2.1 and analysed in the following section. At the core of the concept is circulation, which entails the long-term use of products, components and materials (Ellen MacArthur Foundation 2015). The whole concept of "waste" or "end-of-life" is no longer needed (Ellen MacArthur Foundation 2013b; Kirchherr, Reike, and Hekkert 2017). In the CE, the system is restorative and regenerative (Ellen MacArthur Foundation 2021; Geissdoerfer et al. 2017; Murray, Skene, and Haynes 2017), which means that the impacts on the environment are at a minimum. Furthermore, central to the concept is the R framework. Based on Kirchherr, Reike, and Hekkert (2017), the 3R framework – reduce, reuse, and recycle – is most commonly used. However, the literature also discusses the R10 framework – refuse, rethink, reduce, reuse, repair, refurbish, remanufacture, repurpose, recycle, and recover (Potting et al. 2017). The aforementioned definitions highlight the economic and environmental dimensions while also integrating sustainable development (Kirchherr, Reike, and Hekkert 2017; Korhonen, Honkasalo, and Seppälä 2018; Prieto-Sandoval, Jaca, and Ormazabal 2018), social equity (Kirchherr, Reike, and Hekkert 2017) and human well-being (Murray, Skene, and Haynes 2017) as part of the CE.

Table 2.1 Examples of CE definitions

References	Definition
Desing et al. (2020: 7–8)	"The Circular Economy is a model adopting a resource-based and systemic view, aiming at taking into account all the variables of the system Earth, in order to maintain its viability for human beings. It serves the society to achieve well-being within the physical limits and planetary boundaries. It achieves that through technology and business model innovation, which provide the goods and services required by society, leading to long term economic prosperity. These goods and services are powered by renewable energy and rely on materials which are either renewable through biological processes or can be safely kept in the technosphere, requiring minimum raw material extraction and ensuring safe disposal of inevitable waste and dispersion in the environment. CE builds on and manages the sustainably available resources and optimizes their utilization through minimizing entropy production, slow cycles and resource and energy efficiency".
Ellen MacArthur Foundation (2013a: 7)	"A circular economy is an industrial system that is restorative or regenerative by intention and design…It replaces the 'end-of-life' concept with restoration, shifts towards the use of renewable energy, eliminates the use of toxic chemicals, which impair reuse, and aims for the elimination of waste through the superior design of materials, products, systems, and, within this, business models".
Geissdoerfer et al. (2017: 759)	"A regenerative system in which resource input and waste, emission, and energy leakage are minimised by slowing, closing, and narrowing material and energy loops. This can be achieved through long-lasting design, maintenance, repair, reuse, remanufacturing, refurbishing, and recycling".
Kirchherr, Reike, and Hekkert (2017: 224)	"A circular economy describes an economic system that is based on business models which replace the 'end-of-life' concept with reducing, alternatively reusing, recycling and recovering materials in production/distribution and consumption processes, thus operating at the micro level (products, companies, consumers), meso level (eco-industrial parks) and macro level (city, region, nation and beyond), with the aim to accomplish sustainable development, which implies creating environmental quality, economic prosperity and social equity, to the benefit of current and future generations".
Korhonen, Honkasalo, and Seppälä (2018: 39)	"Circular economy is an economy constructed from societal production-consumption systems that maximizes the service produced from the linear nature-society-nature material and energy throughput flow. This is done by using cyclical materials flows, renewable energy sources and cascading1-type energy flows. Successful circular economy contributes to all the three dimensions of sustainable development. Circular economy limits the throughput flow to a level that nature tolerates and utilises ecosystem cycles in economic cycles by respecting their natural reproduction rates".

(Continued)

Table 2.1 (Continued)

References	Definition
Murray, Skene, and Haynes (2017: 371–377)	"By circular, an economy is envisaged as having no net effect on the environment; rather it restores any damage done in resource acquisition, while ensuring little waste is generated throughout the production process and in the life history of the product… The Circular Economy is an economic model wherein planning, resourcing, procurement, production and reprocessing are designed and managed, as both process and output, to maximize ecosystem functioning and human well-being".
Prieto-Sandoval, Jaca, and Ormazabal (2018: 618)	"Circular economy as an economic system that represents a change of paradigm in the way that human society is interrelated with nature and aims to prevent the depletion of resources, close energy and materials loops, and facilitate sustainable development through its implementation at the micro (enterprises and consumers), meso (economic agents integrated in symbiosis) and macro (city, regions and governments) levels. Attaining this circular model requires cyclical and regenerative environmental innovations in the way society legislates, produces and consumes".

Schöggl, Stumpf, and Baumgartner (2020) show in their review a significant growth in the number of CE studies from 2016 onwards. Our analysis takes a closer look and highlights five aspects of the previous literature (especially Schöggl, Stumpf, and Baumgartner 2020; Merli, Preziosi, and Acampora 2018; Ghisellini, Cialani, and Ulgiati 2016). First, context-wise, the previous studies have focused on either China or Europe (Merli, Preziosi, and Acampora 2018). Second, the previous studies can be divided into three different levels: macro, meso, and micro. Macro-level studies address the CE in a city, region or country (Merli, Preziosi, and Acampora 2018; Ghisellini, Cialani, and Ulgiati 2016). In these studies, the focus has been mostly on its socio-economic dynamics (Schöggl, Stumpf, and Baumgartner 2020; Merli, Preziosi, and Acampora 2018; Ghisellini, Cialani, and Ulgiati 2016). In the meso-level literature, attention is paid to industrial parks (Merli, Preziosi, and Acampora 2018; Ghisellini, Cialani, and Ulgiati 2016). Last, the micro-level articles discuss topics related to individual companies or consumers (Ghisellini, Cialani, and Ulgiati 2016). The third aspect of the previous literature is theme: the focus has been on practical ways to implement the CE, including tools and methods (Schöggl, Stumpf, and Baumgartner 2020; Merli, Preziosi, and Acampora 2018; Ghisellini, Cialani, and Ulgiati 2016). Fourth, despite the fact that there is a large number of previous CE studies, their topics have not varied significantly. Both Schöggl, Stumpf, and Baumgartner (2020) and Merli, Preziosi, and Acampora (2018) notice that most studies have an environmental focus. Schöggl, Stumpf, and Baumgartner (2020) even state that the majority of previous studies focus on recycling, and Merli, Preziosi, and Acampora (2018) mention that waste management is a typical research topic. Fifth, as we have shown above, the CE is often discussed in relation to its economic and environmental dimensions. Many researchers point out that the social and cultural aspects of CE are seldom studied (Schöggl, Stumpf, and Baumgartner 2020; Merli, Preziosi, and Acampora 2018; Geissdoerfer et al. 2017; Kirchherr, Reike, and Hekkert 2017; Murray, Skene, and Haynes 2017), and therefore more focus on them is needed.

Cultural perspectives on the circular economy

In the previous section, we show that existing CE research tends to focus on its economic and environmental dimensions. In contrast, we are interested in its less studied cultural dimension. The cultural dimension covers the aspects of human interactions in a society. Due to this nature, it is often also called the socio-cultural approach. For example, Warner (2010) explains that the socio-cultural dimension consists of changes in the societal demographic structure and its values and beliefs. Brennan and Sisk (2014: 45–46) list "demographic trends, cultural considerations, literacy levels, social infrastructure, consumer confidence, and religious beliefs" under the concept. Yüksel (2012) adds to these items lifestyle and level of education. In the following, we first discuss the concept of culture in general, and then we discuss the CE from the point of view of its socio-cultural dimension.

Traditionally, culture is an umbrella term describing multiple aspects of human behaviour in groups, organisations, and societies. The concept has been under study and debate for decades (or even centuries, if we start with the anthropologies), and authors across different disciplines have offered varying definitions. In a traditional definition proposed by Tylor (according to Bidney 1944), culture is a "complex whole which includes knowledge, beliefs, art, morals, law, customs and any other capabilities and habits by man as a member of society", viewed specifically from an anthropological perspective. Peterson (1979) elaborates that the discussion of culture has focused especially on four elements: norms, values, beliefs, and expressive symbols. While culture is much debated across disciplines (particularly amid the social sciences) (Bennett 2015), there seems to be a somewhat general agreement that culture is created by people as members of societies and communicated, largely via language use (Bidney 1944), but also through other artefacts and values, both visible and invisible, that serve to label our behaviour. We create the culture, and in turn, culture defines us and the ways we live in certain contexts. Through culture we derive assumptions about what is acceptable, justified or morally good. For example, culture can provide legitimate foundations for what is considered a legitimate agreement, solution, or practice in a certain context (see, for example, Park 2005).

Culture is not only multidimensional as a concept but is also able to be perceived at different levels of societies. Culture can be perceived, for example, at the national, regional, industrial, subcultural, organisational, departmental, functional, and team levels, which are always interactively influencing one another (Alvesson and Berg 1992). National cultures are characterised as powerful constructions (Stevenson 1997) marked by complexity (Fang 2015). They are often summarised by simplifications, such as having "collectivist" or "individualistic" orientations. However, within national cultures there are multiple other cultures, such as those of ethnic minorities or regions (Bauman and May 2001). Within organisation studies, the concept of organisational culture is extensively used and debated. In the classic definition, organisational culture involves visible artefacts, partly visible values and underlying, invisible core assumptions (Schein 1990). Organisational culture can powerfully influence the performance of organisations and the human action within them (Warrick 2017), but it can also shape individual experiences (Longman et al. 2018). However, organisational culture is not a coherent whole, but several subcultures have been identified within organisations (Sackmann 1992).

According to the discussion above, culture forcefully influences any processes of change in our societies, including societal and organisational transitions towards the CE. In the existing literature, cultural factors are noted by some authors as a prerequisite for change towards the CE in organisations, though they are not explored in detail. For example, Salvioni and Almici (2020) suggest that the CE requires a change in corporate culture that also engages stakeholders. Also, Kirchherr, Reike, and Hekkert (2017) identify organisational culture as one of the main barriers in transitions towards the CE. Despite these mentions, there is scant explicit research on the cultural element of the CE.

Previous CE studies use the term social CE, as the CE is often connected to sustainability, which includes economic, environmental, and social elements. Therefore, in the following section, we use the term social CE and highlight four aspects of it based on previous studies. First, the social aspects of the CE are studied from a rather limited perspective and are often connected with the economic dimension. For example, Geissdoerfer et al. (2017) state that the only social aspects of the CE that are studied include job creation and efficient tax systems (see also, D'Amato et al. 2017). In addition, Geissdoerfer et al. (2017) and Schöggl, Stumpf, and Baumgartner (2020) point to previous CE studies on shared economy. Taking a closer look, Kirchherr, Reike, and Hekkert (2017) found that such studies often discuss social well-being.

Second, social CE studies tend to focus on consumers. For example, Coderoni and Perito (2020) consider consumers' acceptance of purchasing waste-to-value food. Bovea et al. (2018), in turn, analyse consumers' perception of product labels, along with CE icons and their symbolisation of the CE, while Nainggolan et al. (2019) research consumers and their household waste sorting habits. Although there are some studies on consumers and the CE, Schöggl, Stumpf, and Baumgartner (2020) encourage further research, particularly into consumption patterns and behavioural aspects. This approach is studied in this book (see Chapters 7 and 9).

Third, previous CE reviews have recognised the role of collaboration in CE transitions. Geissdoerfer et al. (2017) highlight that, in order to succeed in implementing the CE, there is a need for stakeholder cooperation. D'Amato et al. (2017) and Schöggl, Stumpf, and Baumgartner (2020) share this focus on cooperation, as they note that the greening supply chain (i.e. collaboration in the supply chain) is a rather common CE research topic. In their recent reviews on the CE, both

Schöggl, Stumpf, and Baumgartner (2020) and Sarja, Onkila, and Mäkelä (2021) raise CE collaboration as an area deserving future research.

Fourth, inspired by the lack of research on the social dimension of the CE, Padilla-Rivera et al. (2021) dig deeper into this field via a literature review, finding 60 studies addressing it. Content-wise, the studies cover three thematic areas: labour practices and decent work; society (including human rights); and product responsibility. Inside these larger themes, the most covered subthemes are: employment; social inclusion; sharing economy; participation and local democracy; and health and safety.

Materials and methods

Interviews

Our empirical material consists of 68 interviews with 71 Finnish CE experts (see Appendix for interview data). The experts represented both public and private organisations (see Table 2.2). The public organisations included ministries, cities, and regional councils. These organisations are the key promoters or stakeholders of the CE in Finland. The titles of the interviewees varied from directors and managers to advisors. Other interviewees represented different types of private manufacturing and service organisations and industrial federations. These organisations are considered CE front-runners in Finland. The size of the private organisations varied from start-ups to large multinational companies. In these organisations, the interviewees were typically CEOs or other directors. Both women and men were interviewed. The duration of the interviews varied between 27 and 100 minutes. The interviews were either conducted face-to-face, often at the interviewee's location, or via the internet using either Zoom or Skype.

The interviews were semi-structured. The themes covered in the interviews included the interviewee's background, CE practices in the interviewee's organisation, and CE implementation in their line of business and in Finland overall. All these themes were covered in every interview, yet the exact questions used in each interview differed due to time limitations and the expertise of the interviewee. All

Table 2.2 Types of organisations the interviewed experts represented

Public/private organisations	Organisation type	Number of interviews
Public organisations	Municipalities and regional councils	8
	Ministries and other governmental organisations	6
Private organisations	Service companies	18
	Manufacturing companies	33
	Industrial federations	3

the interviewees gave their permission to record the interview. Later the interviews were transcribed.

Analysing the data

Transcribed interviews were thematically analysed. The analysis focuses on those parts of the transcripts where the interviewees discuss the cultural aspects of the CE. In practice, this meant multiple rounds of reading and coding the interview transcripts. The program Atlas.ti was used to assist in the coding. Working from the literature on cultural perspectives, we identified the following questions, which we used in the coding:

- What was the role of values?
- What kind of cultures were mentioned?
- What was the role of attitudes?
- What was the role of education and awareness building?
- Which stakeholders were mentioned in relation to the CE?
- What kind of cooperation was performed with the stakeholders?

Results

The analysis of the interviews led us to identify that the discussion of the cultural dimensions of the CE circled around the change needed in order to implement it. The interviewees were unanimous in asserting that our society needs systemic change toward the CE. However, the interviewees also concurred that the change process needs to be made easy. For example, the CE was seen as an opportunity to create ways to ease daily life: thus, complicated attempts to implement the CE will not thrive. Although the need for change was recognised, many of the interviewees talked at length about the resistance to change and even the fear of change among those people less active in CE discussions. The interviewees identified that some individuals and even certain industries are reluctant to change. One example in relation to shared economy was the role of insurance companies. If you share your car and then an accident happens, does your insurance compensate you? These old habits and structures can prevent CE applications from spreading. Furthermore, the change discourse addressed three areas of interest (the change in values and attitudes, raising awareness and knowledge of the CE, and cooperation and solidarity in implementing the CE). These aspects will be discussed in the following section.

First, a change in values and attitudes is needed in order to successfully transition to the CE. On this topic, the interviewees gave mixed answers. Some of the examples mentioned by the interviewees were the emergence of pro-CE and pro-environmental values among their customers and individuals in general. They also listed actions that they themselves undertook in order to promote the CE and highlighted recently adopted ways of acting greener. They also often named their company or the owners of their company as bearing values that promote the implementation of the CE. However, other interviewees mentioned that we still need a larger and wider change of values across society to truly make the CE change, and they perceived the inclusion of everybody in the change as a prerequisite for the success of such transitions. In addition, a few interviewees mentioned that there are industries and companies that do not hold sustainable values. These companies focus on profit maximisation and often see the CE only as an extra cost.

The second theme covers both the awareness and knowledge of the CE. The increase in awareness was mentioned as vital for global CE transitions to occur. Awareness refers to the general understanding of the concept of the CE. On the one hand, the concept is seen as a difficult one. The interviewees were worried about how to make consumers understand the CE and its various elements. On the other hand, they noted that awareness about these issues was already increasing. Nevertheless, they propose that consumers make their purchasing decisions based on emotions and feelings, while businesses base their decisions often only on financial considerations. Often the interviewees gave examples of the ways they raised awareness of the CE among their stakeholders, thus aiming to include them in the change. They participated in different meetings, gatherings, and conferences to give speeches on the CE and to meet new people. In addition, the studied companies were members of different networks, non-profit organisations, and federations (local, national, and international), and each targeted general awareness of CE. Furthermore, the interviewees mentioned that occasionally they still needed to persuade customers that their product or service is a better option in comparison to a non-CE product; this was one way they increased the awareness of the customers. Knowledge of the CE is closely linked with awareness. The main difference between these two concepts is that knowledge is directly connected to the formal education system. In relation to knowledge, the interviewees were rather unanimous. It was generally held that the level of knowledge and education is high in Finland. This fact means, for example, that engineering knowledge and innovations in the CE are prolific. This was considered a potential success factor for CE implementation in Finland.

Finally, cooperation and solidarity were also identified as key success factors. All the interviewees talked at length about cooperation with different parties. It was a generally held perception that one cannot bring change by working alone. Thus, solidarity did not refer here only to a sense of togetherness created through a shared orientation towards changes but also a collective responsibility to create those changes. Four typical approaches to such cooperation were identified. First, many companies cooperated through their supply chains. For example, companies needed cooperation with suppliers in order to obtain recycled material to produce their products. Moreover, they also needed to cooperate with the customers and even their customers' customers in order to close the loop and gather used products to use as new raw materials. Second, the companies cooperated in particular with other companies within their business sector. This teamwork was typical in cases where challenges to the CE were so considerable that it was not possible for an individual company to solve them. In such instances, cooperation also included that industry's federation. Third, new combinations of industries produced new cooperative partners and methods. This was especially true in the case of CE product innovations. Furthermore, cooperation with the public sector was often mentioned. For example, municipalities were often seen as important partners, as the decisions they make can be important in the CE transitions. For example, they may follow CE principles in the development of new urban areas, as was visible in our study for the city of Espoo's Kera or Tampere's Hiedanranta areas. Fourth, the companies and different organisations were eager to take part in different CE projects together with various research institutions. Similarly, the companies offered thesis projects for students from various educational institutions. While the interviewed experts emphasised cooperation, they were also able to name either companies or industries unfamiliar with such cooperation and eager to preserve their current way of operating. Their inclusion in the direction of change was seen as a prerequisite for a societal transition towards the CE.

We have summarised our results in Table 2.3. Each identified cultural aspect is briefly discussed from the point of view of catalysts and obstacles, along with the related inclusion features. In general, catalysts are existing visible changes in values and in awareness, as well as tangible ways of cooperating. Obstacles are largely related to the kind of human actions that have not yet adopted changes towards the CE. From the point of view of inclusion, however, the interviewees did not want to construct opposite views with other stakeholders, but instead

Table 2.3 Summary of the catalysts and obstacles of cultural CE transitions

Identified cultural aspect	Catalysts	Obstacles	Inclusion
Values and attitudes	Change in pro-environmental and pro-CE values is already visible.	The large majority still does not hold very pro-environmental values.	Inclusion of the majority is a prerequisite for the CE transitions, but it is not yet happening.
Raising awareness and knowledge of CE	Awareness of CE is increasing. The interviewed experts acted as change agents and promoted the development of awareness by setting an example. The level of knowledge is high in Finland.	The concept of CE was assessed as being difficult to truly comprehend.	Awareness promotion needs to include larger audiences in CE transitions. Currently, inclusion is not happening yet.
Cooperation and solidarity in CE	Active cooperation with various stakeholders in CE among the interviewees.	Some businesses were seen as reluctant to change and therefore operating business as usual.	Inclusion of reluctant business is a prerequisite for CE transitions, but it is not yet happening.

stressed the need for the inclusion of all the actors within the society in the societal transition towards the CE.

Discussion and conclusions

In this chapter, we analyse the cultural aspects that influence CE implementation in Finland. We demonstrate that while the interviewees discussed multiple issues, all of them were linked to the need for change. The interviewees recognised in particular the need for a change in values and attitudes toward the CE, awareness and knowledge of the CE, and cooperation and solidarity in CE actions in Finnish society.

In our findings, three aspects drew our interest. First, it is both interesting and encouraging that the interviewees were unanimous in recognising the need for change in order to achieve the CE transitions. The unanimousness can be explained by the fact that our interviewees were either the front-runners of CE implementation or active stakeholders in it. However, the interviewees did acknowledge that such change can be frightening, as it fundamentally challenges our ways of living. This admission parallels the findings of Hobson (2020). In turn, some of the interviewees saw here an opportunity to create services that will make everyday life easier.

Second, cooperation between different parties was a dominant theme in the interviewees' answers. Indeed, none claimed that they would be able to solve or implement the CE transition by themselves. Besides traditional ways of cooperating, namely cooperation with their supply chains, the interviewees talked about cooperation with new partners and with other, even very distant business sectors. It is possible that new business opportunities will emerge from this new cooperation.

Third, from the point of view of solidarity and inclusivity, our study offers mixed results. On the one hand, the dominant role of cooperation is encouraging. Our interviewees proposed that we can implement the CE and solve global environmental problems with cooperation based not only on a sense of solidarity born from a shared orientation towards change but also on a sense of collective responsibility for making the change happen. On the other hand, the interviewees were also rather unanimous that there is currently some degree of polarisation in CE implementation. There are people and organisations that promote the CE operating alongside even larger groups of people and organisations maintaining the status quo. The change towards the CE is vitally important for the survival of the planet (Ellen MacArthur Foundation 2013a), such that we do not have time and resources to lose in confrontation. However, numerous industries, organisations, and individuals remain hesitant, unaware or resistant to this change, while their inclusion is a prerequisite for the societal transition to progress.

The main limitation of this study is that is based on one country, given that our aim was to study the phenomenon of the CE specifically in Finland and the interviews were conducted there. Furthermore, Finland aims to be a leading CE country (Finnish Government 2019, 2021), so we believe that it is an interesting case from a global perspective. Going forward, we encourage the study of the cultural CE globally and see the need to conduct interviews in other countries, which will provide comparisons between countries' transitions.

Besides a call to widen the geographical reach of research on the cultural elements of the CE, our study raises three main avenues for future research. First, the use of language used to discuss cultural aspects needs to be further studied. In this study, we only named the different cultural aspects that the interviewees mentioned during the interviews. We did not place particular emphasis on how the interviewees discussed these aspects. Nevertheless, it is important to consider what kinds of words we use to discuss the needed change. For example, are we enthusiastic about the upcoming change and therefore able to encourage others, or are we sceptical or even afraid of change and therefore preventing the change?

The second area for future research is the cooperation needed in CE transitions. In this research, we were only able to scratch the surface of the topic of cooperation. The interviewees talked at length about the cooperative actions that they themselves take and what their own company is doing. This focus opens two avenues: 1) the active role of individuals in promoting change through cooperation and 2) the cooperation between different types of organisations.

A third area of future research is the societal structures that either catalyse or hinder CE transitions. Although we emphasise in our chapter the key role of individuals as decision-makers, we do recognise that individuals always operate in the wider context of society. Occasionally our interviewees mentioned structural issues, such as legislation and various legal requirements or old working habits, as obstacles to CE implementation. These aspects demand a deeper analysis.

Acknowledgements

This work is part of the research project Circular Economy Catalysts: From Innovation to Business Ecosystems (CICAT2025) funded by the Strategic Research Council at the Academy of Finland (320205; 0245896-2018). In addition, we are thankful for the help of our CICAT2025 colleagues (A. Jokinen, J. Kujala, A. Heikkinen, M. Marjamaa, and R. Tapaninaho) in collecting the research data.

Appendix

Table 2.4 The list of interviewees

Date	Code	Organisation/position	Face-to-face/online	Duration	Male/female
8.5.2019	I1	Municipality Development Project/Project Manager	Face-to-face	49 min	F
22.5.2019	I2	Environmental Service Provider A/Business Director	Face-to-face	62 min	M
3.6.2019	I3	City A/R & D Director	Face-to-face	74 min	F
7.6.2019	I4	Government Agency/Programme Director	Online	65 min	F
11.6.2019	I5	Sustainable Development Company/Leading expert	Face-to-face	81 min	M
12.6.2019	I6	IT Company/Sales Director	Face-to-face	49 min	M
13.6.2019	I7	Industry Federation A/Environmental Manager	Face-to-face	61 min	M
13.6.2019	I8	Industry Federation B/Leading Expert	Face-to-face	81 min	F
17.6.2019	I9	City B/Environmental Manager	Face-to-face	65 min	F
18.6.2019	I10	Ministry A/Senior Expert	Face-to-face	77 min	F
18.6.2019	I11	Industry Federation C/Director	Face-to-face	57 min	M
18.6.2019	I12	Ministry B/Special Advisor and Special Advisor	Face-to-face	54 min	F & F
18.6.2019	I13	City C/Sustainable Development Director	Face-to-face	54 min	M
19.6.2019	I14	Municipality Development Project/Project Development Director	Face-to-face	72 min	M
20.6.2019	I15	Regional Council A/Innovation and Future Director	Face-to-face	58 min	M
20.6.2019	I16	Ministry C/Project Director	Online	73 min	F
20.6.2019	I17	Ministry A/Head of a Unit	Online	52 min	F
24.6.2019	I18	Fund/Leading Expert	Online	74 min	M
25.6.2019	I19	City D/Environmental Expert	Face-to-face	72 min	F
8.8.2019	I20	Regional Council B/Project Manager and Development Manager	Face-to-face	78 min	M & F
23.8.2019	I21	Environmental Service Provider B/Circular Economy Specialist	Face-to-face	39 min	F
27.8.2019	I22	Construction Company/Sustainable Business Director	Face-to-face	54 min	M
20.9.2019	I23	Manufacturing company A/ Business Unit Manager	Face-to-face	55 min	F
29.10.2019	I24	Manufacturing Company B/CEO	Face-to-face	28 min	M

(Continued)

Table 2.4 (Continued)

Date	Code	Organisation/position	Face-to-face/online	Duration	Male/female
29.10.2019	I25	Forest Industry Company A/Director of Sustainability	Face-to-face	42 min	F
29.10.2019	I26	Energy Company A/Sales Director	Face-to-face	35 min	M
5.11.2019	I27	Energy Company B/Biorefinery Business Director	Face-to-face	61 min	M
7.11.2019	I28	Forest Industry Company B/Business Unit Director	Face-to-face	71 min	F
11.11.2019	I29	Manufacturing Company C/Chief Marketing Officer	Face-to-face	60 min	M
12.11.2019	I30	Car Sharing Company A/CEO	Face-to-face	49 min	M
13.11.2019	I31	Car Sharing Company B/Marketing and Sales Coordinator	Face-to-face	45 min	F
18.11.2019	I32	Forest Industry Company C/Manager, Environmental Production Support and Responsibility Director, Strategic Partnerships and Technology	Face-to-face	45 min	M & F
19.11.2019	I33	Forest Industry Company D/VP Sustainability	Face-to-face	86 min	M
27.11.2019	I34	Environmental Technology Company A/CEO	Face-to-face	65 min	M
10.12.2019	I35	Waste Management Company/CEO	Face-to-face	62 min	M
11.12.2019	I36	Forest Industry Company C – Subsidiary/Sustainability Expert	Face-to-face	31 min	F
12.12.2019	I37	Consulting Company A/Director, Circular Concepts	Face-to-face	31 min	F
9.1.2020	I38	Financial Company/Investment Director	Face-to-face	54 min	M
17.1.2020	I39	Service Company A/CEO	Face-to-face	68 min	M
4.2.2020	I40	Manufacturing Company D/CEO	Face-to-face	52 min	M
7.2.2020	I41	Non-profit Recycling Company/CEO	Face-to-face	100 min	M
19.2.2020	I42	Biogas Company A/CEO	Face-to-face	54 min	M
24.2.2020	I43	Material Recycling Company/CEO	Face-to-face	35 min	F
25.2.2020	I44	Environmental and Property Maintenance Company/SVP, Corporate Relations	Face-to-face	57 min	M
28.2.2020	I45	Service Company B/CEO	Face-to-face	37 min	M
28.2.2020	I46	Waste Management Company/CEO	Face-to-face	37 min	M
3.3.2020	I47	Online Platform for Second-hand Items A/CEO	Face-to-face	53 min	M

Date	ID	Role	Mode	Duration	Gender
3.3.2020	148	Textile Company A/Senior Vice President, Business Concept Development	Online	65 min	F
4.3.2020	149	Outdoor Textiles and Items/CEO	Face-to-face	48 min	M
5.3.2020	150	Furniture Company/Sustainability Manager	Face-to-face	69 min	F
5.3.2020	151	Interior Design Company/Strategy Director	Face-to-face	50 min	M
5.3.2020	152	Waste Container Manufacturer/Circular Economy Specialist	Face-to-face	45 min	F
9.3.2020	153	Textile Company B/CEO	Online	27 min	F
11.3.2020	154	Civil Engineering Service Company/Chairman of the Board	Online	40 min	M
12.3.2020	155	Biogas Company B/CEO	Face-to-face	68 min	F
12.3.2020	156	Manufacturing Company E/Business Director	Face-to-face	75 min	F
12.3.2020	157	Manufacturing Company F/CEO	Online	40 min	M
16.3.2020	158	Online Platform for Second-hand Items B/Marketing and Communications	Online	47 min	F
17.3.2020	159	Online Platform for Second-hand Items C/CEO	Online	55 min	M
19.3.2020	160	Manufacturing Company G/CEO	Online	69 min	M
26.3.2020	161	Agriculture and Forestry Machine Retailer/CEO	Online	42 min	M
26.3.2020	162	Design Retailer/CEO	Online	47 min	M
26.3.2020	163	Composting Company/Business Manager	Online	27 min	M
31.3.2020	164	Manufacturing Company H/CEO	Online	37 min	M
23.4.2020	165	Textile Company B/CEO	Online	55 min	F
5.5.2020	166	Textile Company C/CEO	Online	41 min	F
18.5.2020	167	Textile Company D/CEO	Online	54 min	M
28.5.2020	168	Manufacturing Company I/Director, Sales and Management	Online	91 min	M

References

Alvesson, Mats, and Per Olof Berg. 1992. *Corporate Culture and Organizational Symbolism*. Berlin: Walter de Gruyter.

Bauman, Zygmunt, and Tim May. 2001. *Thinking Sociologically*. New York: Wiley.

Bennett, Tony. 2015. "Cultural Studies Cultural Studies and the Culture Concept." *Cultural Studies* 29 (4): 546–68. https://doi.org/10.1080/09502386.2014.1000605.

Bidney, David. 1944. "On the Concept of Culture and Some Cultural Fallacies." *American Anthropologist* 46: 30–44. https://doi.org/10.1525/aa.1944.46.1.02a00030.

Bovea, María D., Pilar Quemades-Beltrán, Victoria Pérez-Belis, Pablo Juan, Marta Braulio-Gonzalo, and Valeria Ibáñez-Forés. 2018. "Options for Labelling Circular Products: Icon Design and Consumer Preferences." *Journal of Cleaner Production* 202 (November): 1253–63. https://doi.org/10.1016/j.jclepro.2018.08.180.

Brennan, Linda L., and Faye Sisk. 2014. *Strategic Management: A Practical Guide*. Cognella Academic Publishing, Business Expert Press.

Coderoni, Silvia, and Maria Angela Perito. 2020. "Sustainable Consumption in the Circular Economy. An Analysis of Consumers' Purchase Intentions for Waste-to-Value Food." *Journal of Cleaner Production* 252 (April): 119870. https://doi.org/10.1016/j.jclepro.2019.119870.

D'Amato, D., N. Droste, B. Allen, M. Kettunen, K. Lähtinen, J. Korhonen, P. Leskinen, B. D. Matthies, and A. Toppinen. 2017. "Green, Circular, Bio Economy: A Comparative Analysis of Sustainability Avenues." *Journal of Cleaner Production* 168 (December): 716–34. https://doi.org/10.1016/j.jclepro.2017.09.053.

Desing, Harald, Dunia Brunner, Fabian Takacs, Stéphane Nahrath, Karolin Frankenberger, and Roland Hischier. 2020. "A Circular Economy within the Planetary Boundaries: Towards a Resource-Based, Systemic Approach." *Resources, Conservation and Recycling* 155 (April): 104673. https://doi.org/10.1016/j.resconrec.2019.104673.

Elia, Valerio, Maria Grazia Gnoni, and Fabiana Tornese. 2017. "Measuring Circular Economy Strategies through Index Methods: A Critical Analysis." *Journal of Cleaner Production* 142 (January): 2741–51. https://doi.org/10.1016/j.jclepro.2016.10.196.

Ellen MacArthur Foundation. 2013a. *Towards the Circular Economy: Economic Amd Business Rationale for an Accelerated Transition*. Vol. 1. Ellen MacArthur Foundation.

———. 2013b. *Towards the Circular Economy: Opportunities for the Consumer Goods Sector*. Vol 2.

———. 2015. *Towards a Circular Economy: Business Rationale for an Accelerated Transition*.

———. 2021. "What Is The Circular Economy?" 2021. https://www.ellenmacarthurfoundation.org/circular-economy/what-is-the-circular-economy.

Fang, Tony. 2015. "From 'Onion' to 'Ocean': Paradox and Change in National Cultures." *International Studies of Management & Organization* 35 (4): 71–90. https://doi.org/10.1080/00208825.2005.11043743.

Finnish Government. 2019. "Inclusive and Competent Finland. A Socially, Economically and Ecologically Sustainable Society." https://julkaisut.valtioneuvosto.fi/handle/10024/161664.

———. 2021. "New Directions: The Strategic Programme to Promote a Circular Economy." https://valtioneuvosto.fi.

Geissdoerfer, Martin, Paulo Savaget, Nancy M.P. Bocken, and Erik Jan Hultink. 2017. "The Circular Economy: A New Sustainability Paradigm?" *Journal of Cleaner Production* 143 (February): 757–68. https://doi.org/10.1016/j.jclepro.2016.12.048.

Ghisellini, Patrizia, Catia Cialani, and Sergio Ulgiati. 2016. "A Review on Circular Economy: The Expected Transition to a Balanced Interplay of Environmental and Economic Systems." *Journal of Cleaner Production* 114 (February): 11–32. https://doi.org/10.1016/j.jclepro.2015.09.007.

Gregson, Nicky, Mike Crang, Sara Fuller, and Helen Holmes. 2015. "Economy and Society Interrogating the Circular Economy: The Moral Economy of Resource Recovery in the EU." *Economy and Society* 44 (2): 218–43. https://doi.org/10.1080/03085147.2015.1013353.

Hobson, Kersty. 2020. "'Small Stories of Closing Loops': Social Circularity and the Everyday Circular Economy." *Climatic Change* 163 (1): 99–116. https://doi.org/10.1007/s10584-019-02480-z.

Kirchherr, Julian, Denise Reike, and Marko Hekkert. 2017. "Conceptualizing the Circular Economy: An Analysis of 114 Definitions." *Resources, Conservation and Recycling* 127 (December): 221–32. https://doi.org/10.1016/J.RESCONREC.2017.09.005.

Korhonen, Jouni, Antero Honkasalo, and Jyri Seppälä. 2018. "Circular Economy: The Concept and Its Limitations." *Ecological Economics* 143 (January): 37–46. https://doi.org/10.1016/j.ecolecon.2017.06.041.

Longman, Karen, Jessica Daniels, Debbie Lamm Bray, and Wendy Liddell. 2018. "How Organizational Culture Shapes Women's Leadership Experiences." *Administrative Sciences* 8: 1–16. https://doi.org/10.3390/admsci8020008.

Merli, Roberto, Michele Preziosi, and Alessia Acampora. 2018. "How Do Scholars Approach the Circular Economy? A Systematic Literature Review." *Journal of Cleaner Production.* Elsevier Ltd. https://doi.org/10.1016/j.jclepro.2017.12.112.

Murray, Alan, Keith Skene, and Kathryn Haynes. 2017. "The Circular Economy: An Interdisciplinary Exploration of the Concept and Application in a Global Context." *Journal of Business Ethics* 140: 369–80. https://doi.org/10.1007/s10551-015-2693-2.

Nainggolan, Doan, Anders Branth Pedersen, Sinne Smed, Kahsay Haile Zemo, Berit Hasler, and Mette Termansen. 2019. "Consumers in a Circular Economy: Economic Analysis of Household Waste Sorting Behaviour." *Ecological Economics* 166 (December): 106402. https://doi.org/10.1016/j.ecolecon.2019.106402.

Padilla-Rivera, Alejandro, Breno Barros Telles do Carmo, Gabriella Arcese, and Nicolas Merveille. 2021. "Social Circular Economy Indicators: Selection through Fuzzy Delphi Method." *Sustainable Production and Consumption* 26 (April): 101–10. https://doi.org/10.1016/J.SPC.2020.09.015.

Park, Yoosun. 2005. "Culture as Deficit: A Critical Discourse Analysis on the Concept of Culture in Contemporary Social Work Discourse." *Journal of Sociology and Social Welfare* 32: 11–33. https://doi.org/NA.

Peterson, Richard A. 1979. "Revitalizing the Culture Concept." *Annual Review of Sociology* 5: 137–66. https://doi.org/10.1146/annurev.so.05.080179.001033.

Potting, José, Marko Hekkert, Ernst Worrell, and Aldert Hanemaaijer. 2017. *Circular Economy: Measuring Innovation in the Product Chain.* The Hague: PBL Netherlands Environmental Assessment Agency.

Prieto-Sandoval, Vanessa, Carmen Jaca, and Marta Ormazabal. 2018. "Towards a Consensus on the Circular Economy." *Journal of Cleaner Production* 179 (April): 605–15. https://doi.org/10.1016/j.jclepro.2017.12.224.

Sackmann, Sonja A. 1992. "Culture and Subcultures: An Analysis of Organizational Knowledge." *Administrative Science Quarterly* 37 (1): 140–61. https://doi.org/10.2307/2393536.

Salvioni, Daniela M., and Alex Almici. 2020. "Transitioning Toward a Circular Economy: The Impact of Stakeholder Engagement on Sustainability Culture." *Sustainability* 12: 8641. https://doi.org/10.3390/su12208641.

Sarja, Milla, Tiina Onkila, and Marileena Mäkelä. 2021. "A Systematic Literature Review of the Transition to the Circular Economy in Business Organizations: Obstacles, Catalysts and Ambivalences." *Journal of Cleaner Production.* Elsevier Ltd. https://doi .org/10.1016/j.jclepro.2020.125492.

Sauvé, Sébastien, Sophie Bernard, and Pamela Sloan. 2016. "Environmental Sciences, Sustainable Development and Circular Economy: Alternative Concepts for Trans-Disciplinary Research." *Environmental Development* 17 (January): 48–56. https://doi .org/10.1016/j.envdev.2015.09.002.

Schein, Edgar H. 1990. "Organizational Culture." *American Psychologist* 45 (2): 109–19. https://doi.org/10.1037/0003-066X.45.2.109.

Schöggl, Josef Peter, Lukas Stumpf, and Rupert J. Baumgartner. 2020. "The Narrative of Sustainability and Circular Economy: A Longitudinal Review of Two Decades of Research." *Resources, Conservation and Recycling.* Elsevier B.V. https://doi.org/10 .1016/j.resconrec.2020.105073.

Stevenson, Nick. 1997. "Globalization, National Cultures and Cultural Citizenship." *The Sociological Quarterly* 38 (1): 41–66. https://doi.org/10.1111/j.1533-8525.1997 .tb02339.x.

Warner, Alfred G. 2010. *Strategic Analysis and Choice: A Structured Approach.* New York: Business Expert Press.

Warrick, D. D. 2017. "What Leaders Need to Know about Organizational Culture." *Business Horizons* 60 (3): 395–404. https://doi.org/10.1016/j.bushor.2017.01.011.

Yüksel, Ihsan. 2012. "Developing a Multi-Criteria Decision Making Model for PESTEL Analysis." *International Journal of Business and Management* 7 (24): 52–66. https:// doi.org/10.5539/ijbm.v7n24p52.

3 Filling the social gap in the circular economy

How can the solidarity economy contribute to urban circularity?

Raysa França, Erkki-Jussi Nylén, Ari Jokinen, and Pekka Jokinen

Introduction

The circular economy (CE) has recently become one of the leading environmental policy concepts (Kirchherr et al. 2017; Lazarevic and Valve 2017; McDowall et al. 2017; Ghisellini et al. 2016), but researchers have questioned its ability to generate change towards sustainability (e.g., Fitch-Roy et al. 2020; Hobson 2020; Korhonen et al. 2018). Various researchers have pointed out that ignorance of the social and cultural aspects of sustainability is a major flaw in the current CE framework (e.g., Corvellec et al. 2020; Schöggl et al. 2020; Korhonen et al. 2018; Geissdoerfer et al. 2017; Moreau et al. 2017; Murray et al. 2017). This lack of social and cultural aspects is particularly pressing in cities, where the legitimacy of CE development needs to be recognised by residents and other stakeholder groups to become a real game-changer for urban sustainability (Nylén et al. 2021; Kębłowski et al. 2020; Williams 2019; Prendeville et al. 2018).

Over 50% of the global population currently lives in urban areas, and the global urban population is expected to increase by 1.5 times by 2045 (World Bank 2020). Besides being the home of the majority of the global population, cities are also centres of major social and environmental problems experienced around the globe (Bartone 1991; Rees 1997; Angotti 1996; Romano and Zullo 2014; Nilsson 2006; Fistola 2011).

The purpose of this chapter is to examine whether a solidarity approach could fulfil the "social gap" in CE. The solidarity approach is represented by the concept of social and solidarity economy (SSE). Moreau et al. (2017) establish a possible parallel between CE and solidarity economy, but a connection between solidarity and circular approaches in urban development is rarely investigated. Gutberlet et al. (2017) discuss joining SSE with CE, but they focus on the role of waste pickers in the Global South and not on urban issues specifically.

The main question we address is: *how can the ideas of SSE enrich social sustainability in urban CE?* SSE and CE are divergent approaches that come from different traditions of research and practice. However, they both argue that the current economic system should be made more sustainable. They discuss the

DOI: 10.4324/9781003255246-3

problems and approach solutions to sustainability in differing ways. CE has so far been mostly concerned with material throughput, stocks, flows of goods, and resources and waste in the economy (Savini 2019; Geissdoerfer et al. 2017; Haas et al. 2015). In contrast, SSE encompasses a series of practices and movements that offer alternatives to the current capitalist system, and it focuses on the principles of cooperation, solidarity, and democracy (Singer 2000).

Our study is based on a literature review of the ideas of CE and SSE followed by one in-depth and two complementary cases of CE and SSE in urban policies in Latin America and Europe. The cases were selected because they illustrate how both concepts can be practised in policymaking and urban development. The overall aim is to reflect and learn from these case examples how to better integrate social and cultural dimensions into CE in the urban context. Europe and Latin America were chosen to utilise the maximum variation of potential case examples. SSE is more established in Latin America (North and Scott Cato 2017; Saguier and Brent 2017), whereas CE is more established in Europe (Fitch-Roy et al. 2020; McDowall et al. 2017).

The ideas of economies

Modern economies are complex systems that in many cases exceed fulfilling basic human needs. Simultaneously, they produce environmental and social problems, which have resulted in large numbers of people living in poverty (Hahnel 2014; Brown 2013). Currently, capitalism is the prevailing organisational theory for economies and has expanded to almost the entire planet. It has achieved impressive results in terms of productivity, efficiency, technological development, and creation of useful goods, but it has failed to decrease gender, class, and country disparities (Cárdenas et al. 2016; Foster 1992). Due to the various environmental and social problems associated with capitalism, different ideas and policy concepts have been generated for how to organise economies (e.g., Raworth 2017; Loiseau et al. 2016).

Sustainable development is the most widespread policy concept for fixing the problems of modern economies (Langhelle 2017). It has enormous political clout: in 2015, the United Nations announced a new agenda for sustainable development with 17 goals (SDGs) to be met by 2030 (United Nations 2015). However, according to a report by independent scientists, no country is on track to meet all of the SDGs by 2030 (United Nations 2019; Sachs et al. 2020). Moyer and Hedden (2020) conclude that the world is not close to achieving many of these SDGs, and some countries are even at risk of achieving none of them.

In general, policy concepts are tools for change as they can address problems and suggest ways to solve them. Different concepts can have kinship if they aim to tackle similar types of problems. However, these concepts differ in how they specify and scale the problems and solutions as well as the words used to describe them. The differing articulations of problems and solutions mean that policy concepts are often contested (Béland 2019; Meadowcroft and Fiorino 2017; Kenis and Lievens 2014; Pollitt and Hupe 2011; Gallie 1956). For example, the concept of sharing economy (Weili and Khan 2020) refers to having or using common underutilised assets through non-monetary or monetary exchange. This implies decreasing the production of new assets and thus reducing the environmental impacts; however, there is evidence that the opposite can occur (Hobson 2020).

Meadowcroft and Fiorino (2017) view policy concepts as innovations that diffuse among the prominent actors and to relevant policy documents. It is more likely that conceptual innovation will achieve success in policymaking if the policy concept is not "too alien to existing discursive patterns and dominant understanding of the way 'the world works'" (Meadowcroft and Fiorino 2017: 11). A prime example is the concept of green capitalism. It states that the technologies that are harmful to the environment can be replaced, capitalist growth can continue within the planet's boundaries, and the greening of the economy will result in access to new markets and profits (Harris 2013). Green capitalism has emerged as a proposal amidst the paradigm of green modernisation (Tienhaara 2014). Drawing from the critical political economy, Brand and Wissen (2015) argue that green capitalism is the culmination of the green economy strategies.

Sullivan (2009) finds that the vocabulary of green capitalism assigns to nature the role of a service provider. Tienhaara (2014) concludes that green capitalism is not a single project but comprises multiple varieties, the common aspect of which is a belief that investing in green sectors can boost employment. Yet, there are different views regarding the acceptable level of regulation and intervention by the state (Tienhaara 2014). Thiele (2019) sees decoupling as a macro strategy of green capitalism as it focuses on technological fixes that can increase eco-efficiency and detach environmental burden from economic growth.

Next, we review two major concepts that are topical in political discussion and policymaking: CE and SSE. Before we move to empirical analysis, we will summarise key elements of green capitalism, the CE, and the SSE, and analyse the differences between them.

Circular economy

Various actors such as non-governmental organisations (NGOs), civil servants, politicians, researchers, and think tanks have generated policy concepts, but the success of these concepts depends on their diffusion. In that sense, the Ellen MacArthur Foundation has been extremely successful with its CE concept. The idea of CE as a policy objective has spread efficiently into influential organisations, such as the European Union (Fitch-Roy et al. 2020; McDowall et al. 2017). The CE narrative revolves around the idea that environmental problems can be tackled with CE solutions and, simultaneously, that these solutions are profitable for business (Nylén 2019; Lazarevic and Valve 2017).

By its very nature, CE is an abstract concept. Multiple different CE definitions exist (Kirchherr et al. 2017), but a general objective can be recognised: it is a closed-loop economy. The CE objective starts by labelling the prevailing model as a linear economy, which functions on the basis of linear throughput: natural virgin resources are extracted, produced as goods, and discarded after consumption as waste. CE promotes various solutions for closing the loop and decreasing the amount of natural virgin resources extracted and wastes generated (Nylén et al. 2021; Geissdoerfer et al. 2017; Ghisellini et al. 2016). As such, CE is a meta concept (Meadowcroft and Fiorino 2017) that gathers these solutions together under its general objective and thus generates momentum towards change.

As CE has become topical in policymaking, it has generated a great deal of research interest as well as criticism. Among the key endorsements for CE are the new business opportunities it is believed to create. However, CE business models are not as easily profitable as the rhetoric might lead one to believe (Nylén 2019; Gregson et al. 2015). Moreover, several researchers have raised the question of whether various CE practices, such as "take back and recycle" schemes, actually increase the amount of material consumed (e.g., Hobson 2020; Valenzuela and Böhm 2017; Hobson and Lynch 2016). According to Korhonen et al. (2018), if the CE discourse does not begin to emphasise the need to decrease the amount of consumption, then it is in danger of becoming more of a technological fix than an actual idea for remodelling the economy. Various scholars have also pointed out the lack of social dimension in CE thinking and policies (e.g., Corvellec et al. 2020; Fratini et al. 2019; Geissdoerfer et al. 2017; Moreau et al. 2017).

Despite these criticisms, proponents still believe that CE can transform economies. However, the changes it proposes are still very much in development, and criticism can affect the pathways that CE policymaking takes. The urban context is an important example. Research on circular cities shows that CE can be fitted to city-level policies, but technological and economical aspects dominate its implementation. This is a serious drawback as a multitude of lifestyles, livelihoods, and social practices exist in cities (Nylén et al. 2021; Kębłowski et al. 2020; Williams 2019; Prendeville et al. 2018). Likewise, Fratini et al. (2019) highlight that attention is rarely given to the role of citizens, authorities, and institutions or to social identities in CE studies. Further, they note that few studies have addressed the relationships between CE and the institutional arrangements underpinning urban transformations. Finally, Moreau et al. (2017) argue that noncompetitive and non-for-profit initiatives and enterprises have not been fully explored in the CE literature.

Social solidarity economy

The SSE concept was originally not introduced by a single influential NGO or think tank. It derives from civic movements and networks; volunteer action; and ideas of solidarity, cooperativism, and democracy in order to fight poverty, inequality, and unsustainable forms of production and consumption, among other socio-economic issues (Moreau et al. 2017; Singer 2000; Cruz 2007). As such, SSE has diverse roots in the concepts of social economy and solidarity economy. The latter began in France and South America in the early 1980s, while the concept of the social economy dates back to discussions of cooperatives or mutual associations from the mid-nineteenth century onwards. The combination of social economy and solidarity economy into SSE took place at the turn of the millennium, as did the founding of the Intercontinental Network for the Promotion of the Social Solidarity Economy (Poirier 2014). As such, Moreau et al. (2017) view SSE as both a practice and a social movement.

As a policy concept, SSE is diverse and somewhat ambivalent. In its perhaps most revolutionary form originating from the solidarity economy, SSE would not aim to achieve cooperativism under the capitalist system but rather to offer a set of economic initiatives of associative character based on an ethics of egalitarianism and diversity. It would be based not only in different structures and institutions but also different ethics, morality, and values (Dacheux and Goujon 2011; Calvo and Morales 2017). In a similar vein, Singer (2000) argues that the solidarity economy should be conceptualised as a mode of production based on the principles of collective or associated ownership of capital and the right to individual freedom. These principles would be applied by a single class of workers who share capital in cooperatives or economic societies, which would result in solidarity and equality. For Singer (2000), reproduction in SSE would be carried out by state mechanisms for income redistribution.

In the literature, SSE authors generally manifest an anti-capitalistic line of thought in different ways. Razeto (1999) understands the solidarity economy as a transforming force inside the mainstream system. For Laville (2009), it comprises a hybrid of volunteer action, market activities by professionals, and activities financed by state subsidies. In that, the diversity of SSE goes beyond orienting its position to capitalism and the transformation of the capitalist economic system. According to Saguier and Brent (2017), SSE in Latin America is a paradigm that values work more than capital. Morandeira-Arca et al. (2021) point out that the vocabulary of SSE also includes the appreciation of a plural economy, including the principles of exchange, redistribution, and non-monetary reciprocity. Furthermore, for Miller (2010), the solidarity economy is an open movement that is always subject to innovation and constant development.

All in all, SSE is a broad and diverse concept that entails a plethora of ideas, aims, and practices that have emerged from the grassroots level of action. As the concept of SSE has become more popular, it has gained attention at the policy level. However, this has not occurred without complications. For example, Saguier and Brent (2017) analyse how the SSE agenda has been transmitted to the Union of South American Nations and Southern Common Market. The adoption of SSE concepts by regional policy has led to a narrow conception of SSE focused primarily on employment. SSE's critiques of economic growth, consumerism, materialism, and extractivism, however, have been neglected. SSE has also been pointed out in the literature for its promising potential in solving issues of poverty and social exclusion (Lee 2019).

Even though SSE seems like a project of civilisation or a human-first approach to a just and socially sustainable economy, its environmental aspects have gained attention in both SSE social movements and policymaking. According to Miller (2010), in a solidarity economy, there is a moral responsibility regarding basic raw materials and the natural processes that sustain life on the planet. One example of the environmental dimension in SSE comes from waste pickers in Brazil,

Table 3.1 Key elements of green capitalism, CE, and SSE

Aspects	Green capitalism	Circular economy	Social solidarity economy
Worldview	Free market (neoliberal economy), decoupling	Green growth, dematerialisation	Degrowth, solidarity, and reciprocity
Role of citizens	Consumers	Consumers, producers, service providers, owners	Citizen-led worker movements
Problems to be solved	Loss-making policies, market failures, environmentally damaging technologies, growth that does not respect the planet's boundaries (Harris 2013)	Linear material and energy flows; environmental pressures that should be decoupled from economic growth (Ghisellini et al. 2016)	Profit over people, loss of ecological diversity, climate change, unemployment (Singer 2000)
Locus for action and attitude toward globalisation	Led by national and international institutions (World Bank, IMF, etc.), tendency towards technologically mediated globalisation	Micro-level: customers, eco-design, cleaner production; meso-level: industry symbiosis; macro-level: cities; national level: policy (Ghisellini et al. 2016)	Self-managed entrepreneurs and local governments, against globalised capitalism or transforming capitalism
Policy instruments	Market-based mechanisms (e.g., carbon tax), carbon trading	Promotion of eco-industrial symbiosis, taxation of polluting practices, fiscal incentives, public procurement rules, innovation vouchers	Distribution, social/community currency, shared decision making, collective ownership of land, community-based consumption, micro-crediting, public banks, decentralisation of the governance of environmental goods

Note: table adapted from Quiroz-Niño and Murga-Menoyo (2017) and Mochizuki and Yarime (2016).

who organise their activities through solidarity and cooperativism. Sabedot and Neto (2017), for example, found that waste pickers recycled 2.3 to 5.7 times more than the formal waste collection of the municipality of Esteio.

Table 3.1 condenses and summarises the key elements of the different ideas of CE, green capitalism, and SSE.

Empirical cases

Research design

To analyze where CE could most productively benefit from the principles of SSE in urban development settings, we applied a case study strategy (Eisenhardt 1989). Taking inspiration from the concept of windows of opportunity (Kingdon 1995), we predicted that collaborative actors working on establishing SSE or CE can utilise multiple opportunity streams for social sustainability depending on strategic vision (problem stream), implementation practices (solution stream), and the arrangements of urban governance (political stream). We thus selected three cases to explore general conditions for social sustainability (Table 3.2).

Our main research case is Hiedanranta, which is a district-scale urban living lab and urban development project of the Finnish city of Tampere. Hiedanranta is a unique case of circular city development. During the first years of the process, the strong citizen activity in bottom-up demonstrations and experiments (Turku et al. 2021) resulted in spontaneous manifestations of social sustainability similar to SSE. To gain a more comprehensive view of potentially productive links between CE and SSE in urban development, we explored suitable SSE cases in Portuguese- and Spanish-speaking regions where SSE principles are adopted (perhaps most widely in the world) and selected two to complement our research setting. The Strategy for Resilience of Montevideo in Uruguay is exemplary of a resilience-centred strategy anchored in both solidarity and circularity aspects and offering a concrete example of how considering both economic ideas could look in policymaking. The Strategy for Social and Solidarity Economy of Madrid in Spain was chosen due to the centrality of solidarity concepts in steering the municipality's economic development. The strong reliance on social discourses,

Table 3.2 Case description

Aspects	Tampere strategy for CE	Montevideo strategy for resilience	Madrid strategy for SSE
Scale	City district	City	City
Vision	Building a new city district based on CE principles	Using resilience framework for promoting SSE and CE	Making SSE comprehensive and growing
Implementation	Urban living lab for multi-actor co-production of CE solutions	Focus on territorial ordering and territory regulation as well as inclusive SSE principles	Economic development, jobs, CE, cooperatives; participation and open forums
Mode of urban governance	Platform governance	Top-down governance	Hybrid or bottom-linked governance
Data	Strategy document, case studies	Strategy document	Strategy document

such as those surrounding inequality and employment, illustrates how CE could be implemented as a social process.

For the data, we examined the following policy documents using content analysis (Creswell 2013): *Hiedanranta Structure Plan 2017*, *The City of Madrid's SSE Strategy 2018–2025*, *Strategy for Resilience in Montevideo 2018*, and *Two Years of Strategy of Resilience in Montevideo*, with occasional analysis of supporting documents. To analyse actions in the Hiedanranta urban living lab, we utilised results from previous case studies based on participatory action research (e.g., Alatalo et al. 2017; Leino and Puumala 2021; Särkilahti et al. 2021; Turku et al. 2021) and several reports published by the city of Tampere.

Case analysis

Hiedanranta Urban Living Lab in Tampere

Tampere in Finland has 235,000 inhabitants and is the fastest growing city in the country. It hosts one of the biggest urban experiments, the Hiedanranta project, which uses CE and smart city development as a guiding principle for urban regeneration. Hiedanranta is a former industrial area that the city is developing into a new urban district that will have 25,000 residents and 10,000 workplaces.

The city began the project by opening the area to city dwellers, civic groups, businesses, and university researchers for CE experimentation. In three years (2016–2019), Hiedanranta developed into an indicative example of platform governance with an active urban living lab hosting approximately 40 experiments and research and development projects. During this period, the city primarily wanted people to create, experiment, and pilot new CE-related ideas for the new urban district and only loosely steered the activities in Hiedanranta at this stage. Rather, the city was an enabler and partner in co-production (Särkilahti et al. 2021).

Consequently, innovative citizen activities (including hundreds of people in total) took place during that period and resulted in the spontaneous development of activities focused on CE in addition to features of social sustainability identified in the SSE literature and movements. This can be seen, for instance, in the entrepreneurial spirit that dominates among cultural actors, which represent the largest citizen group operating out of the old factory buildings in Hiedanranta. Artisans, artists, and entrepreneurs have created a community of more than 20 people, and also some other cultural actors have intentionally found jobs via a hobby started in Hiedanranta. The cultural actors have steered their livelihoods in ways that align with the social economy literature. They 1) seek alternatives to prevalent working life, production, and consumption; 2) emphasise community welfare, cultural values, low-threshold participatory possibilities, community, and collaboration; 3) combine business and social goals in their activities to support the values of a community economy; 4) have a shared feeling of cooperation and personal connections fostered through employment; and 5) draw inspiration from CE by using waste material in their products or providing reparation services (Turku et al. 2021). We read these findings as hidden features of SSE, suggesting that in some circumstances, the ideas of CE and SSE can co-develop and cross-fertilise each other in urban development in ways that support social sustainability.

Considering the evolution of the Hiedanranta urban living lab, cultural actors and other civic groups have not created an isolated niche in the area. Instead, functional links supporting social sustainability have developed between several multi-actor groups. For instance, citizens, companies, and university researchers have participated in many co-creation and co-production projects carried out by the city. This interaction partly contradicts the common understanding that social innovations and technical innovations are distinct activities in urban development. However, although many citizens, cultural actors, and other civic groups have participated in shared projects, they have been uncertain about the collective goals needed to ensure community development during the district's development. They have criticised big development projects that emphasised technology and business. Indeed, the current citizen activities are at risk of ending when construction begins in the area to turn it into a new residential district.

Strategy for Resilience *in Montevideo*

Montevideo is a port city and the capital of Uruguay. Its estimated population of 1.4 million people corresponds to half of the country's total population. The *Strategy for Resilience*, published in 2018, aims to respond to the growing urbanisation of the city, which continues despite the fact that population growth has stabilised. The strategy is mobilised and financed by the Rockefeller Foundation as part of the 100 Resilient Cities network and led by the Department of Urban Planning, with other municipal bodies also involved.

The strategy deals with the following topics: expansion of the urban area and territorial development model; social, economic, and territorial inequality; mobility; transportation; environmental sustainability; waste management; and climate change and climate risk management for coastal territories. When it comes to the economic pillars of the strategy, the goal is to transition Uruguay's economy toward a circular, social, and solidarity economy.

With respect to the economic goals, Montevideo's strategy for resilience does not point to mechanisms commonly associated with SSE (e.g., forms of solidarity exchange and ownership, progressive/distributive taxations, shared ownership, and grassroots' accountability). Thus, the strategies and actions employed by Montevideo's plan only provide a limited picture of SSE objectives for advancing equality and promoting decent employment. Examples of goals related to solidarity include inclusion and equality through the cultural sector as well as reversing spatial segregation in Montevideo. In this sense, the strategy's focus on resilience might be one of the explanatory reasons behind the dissolution of SSE. The strategy views SSE and CE as better equipped to deal with the global challenges that will generate uncertainty in the future. Furthermore, the strategy aims at solving inequality to enable citizens to become key actors in the identification of risks and, therefore, enablers of resilience.

The concept of resilience contains inherent plasticity, which allows multiple interpretations of resilience-building to be adopted, though these do not necessarily follow the radicalism of SSE. Urban resilience, for example, has been characterised as intangible and fragmented (Wardekker et al. 2020). Environmental anthropologists also argue that a focus on resilience avoids an examination of the socio-economic structures behind inequality (Brightman and Lewis 2017). Similarly, Wardekker et al. (2020) argue that practitioners may connect different stakeholders and projects under the umbrella of resilience without necessarily offering interventions that would improve the resilience of the city.

As for the city's specific policy on CE, it appears to be following an approach based on regulation and prohibition. When it comes to the relation of private companies and the public sector, for example, the strategy transfers the cost of waste to construction companies in order to force them to consider creative approaches to dealing with waste. However, Montevideo's strategy fails to provide a policy framework by which solidarity and circularity are considered in a complementary manner. The strategy for circularity, for example, addresses the financing and encouragement of new businesses without citing or taking into consideration solidarity tools and cooperative-based enterprises. Policy instruments of Montevideo's strategy include, among others, taxation of polluting practices, recovery of abandoned areas, public mobility, and construction of new paths for transportation.

SSE in Madrid

Madrid, the capital of Spain, has established the *Municipal Strategy for Social and Solidarity Economy of the City of Madrid* covering the period of 2018–2025. The locus of the strategy is local districts and neighbourhoods, and the strategy is led by the Office for Economic Development. The strategy pursues four objectives: 1) to establish lines of action to achieve SSE centrality in the economic planning of the city, 2) to approximate the SSE economic reality to Madrid's citizens, 3) to territorialise SSE, and 4) to diversify and strengthen the productive fabric of SSE in the city.

The aim of the strategy is to implement SSE in all areas and policies of the municipality, with a special focus on economic development and employment initiatives. Nevertheless, the proposed solutions primarily address SSE as a separate initiative, for example, by determining the creation and curation of spaces dedicated exclusively to SSE. It is unclear, however, how these initiatives would contribute to SSE being integrated into existing businesses and public projects. The strategy has ambitious goals: to create resilience through co-creation, equality, solidarity, and CE while boosting innovation. The bottom-linked governance reflected in the strategy is probably suitable for the urban regime change occurring in Madrid (see Medina-García et al. 2021).

Madrid's strategy brings up a relevant concept concerning the idea of growth: the growth of SSE should not be confused with economic growth (Esber Elias 2019). Whereas in green capitalism growth is associated with the generation of wealth, Madrid's strategy refers to the growth of SSE as a transformative tool capable of reaching traditional businesses and increasing the number of solidarity initiatives in the city. In this, the growth of the solidarity economy would be measured by the number of solidarity enterprises and the increase in decent employment rather than by urban competitiveness or economic growth, which are commonly used as urban development performance indicators.

In Madrid, SSE is seen as a tool to promote inclusion and as a way to build an alternative model of the economy to the current system. In Madrid's SSE, a "human-first" type of economy, democratic and horizontal models for management, and social initiatives and cooperatives are preferred over the "businesses as usual". When it comes to the role of citizens, Madrid's strategy envisions an open forum for SSE deliberation and participation and views citizens as workers, by which a transformation in citizenship can happen through work. Policy instruments include financing SSE initiatives, dissemination through formal education, inclusion of vulnerable people who have difficulties entering the formal job market, research and development, clustering, and territorial centres for SSE innovation.

Broadening the opportunities for social sustainability

Our theoretical and empirical findings show, first, that central principles of social sustainability are inherent in SSE, but the realisation of these principles, for instance in Montevideo and Madrid, depends on several factors. Second, while previous research indicates a severe lack of social sustainability in CE, the Hiedanranta case provides some evidence to the contrary. The manifestations of social sustainability in Hiedanranta, bearing substantial similarities with those of SSE, required favourable circumstances to emerge.

Consequently, both in SSE and CE, a successful realisation of the principles of social sustainability depends crucially on favourable circumstances resulting from convergence in opportunity streams (strategic vision, implementation practices, and urban governance) and skilful promoters who utilise these circumstances. This is how an opportunity window for social sustainability opens (see Kingdon 1995).

Table 3.3 summarises the ideas of social sustainability that we found both in the SSE literature and the analysed cases. The table also shows empirical examples of the SSE ideas (including the ones that have emerged in Hiedanranta) that could contribute to the implementation of CE in urban contexts and be used for advancing urban sustainability. We found that SSE can contribute to CE and urban sustainability in major aspects, such as stakeholder interaction, barriers to implementation, employment, democracy, business models, cultural heritage, education, and resilience.

In CE, the social dimension can be an important factor in determining sustainability in the long term (Moreau et al. 2017; Padilla-Rivera et al. 2020). Drawing from our findings, we propose that the ideas of SSE can enrich social sustainability in CE in cities in several ways. However, this can take place only in suitable circumstances defined by a strategic vision, implementation practices, and urban

Table 3.3 Ideas of SSE contributing to social sustainability in CE

SSE ideas	*Case*	*Contribution of SSE to CE and urban sustainability*
Economy as a social process	Madrid and Montevideo: SSE encourages territorial equality, decent employment, gender equality, and inclusiveness. Hiedanranta: Cultural actors exercise practices of social economy and value jobs.	SSE ideas and practices can facilitate the overcoming of the social barriers of CE implementation, such as political disputes on distribution or unequal access to resources.
Role of citizenship and community participation	Madrid: An open forum for SSE is provided. Hiedanranta: The whole urban living lab was opened for civic experimentation and co-production between the city, citizens, businesses, and researchers.	SSE may offer ideas on how to increase stakeholder interaction to aid CE implementation (Winans et al. 2017). SSE can bring a needed focus on communities (Lee 2019).
Instruments to promote inclusion	Madrid: SSE includes vulnerable populations who cannot access the formal job market. Hiedanranta: Single bottom-up initiatives are received from anyone.	SSE and CE can provide instruments for social inclusion and thereby promote social sustainability.
Social innovation	Madrid and Montevideo: Innovative bottom-up initiatives are considered in SSE. Hiedanranta: Cultural actors and other groups create a community supporting social innovations related to their entrepreneurial ideas, sometimes working with other actor groups operating in the area.	Social innovation and experiments are often key to successful CE implementation (Bulkeley et al. 2016). Art contributes to the imagination of sustainable futures.
Alternative business models	Madrid: SSE contributes to the realisation of cooperative business models. Hiedanranta: Social entrepreneurship among cultural actors is developed.	SSE can inform and help design much-needed new business models which contribute to sustainability transitions (Geissdoerfer et al. 2018)
Role of cultural goods	Montevideo: Preservation of culturally relevant sites strengthens resilience. Hiedanranta: Activities of cultural actors parallel socio-technical urban development but are not necessarily balanced with it in the long term. Artisans want to design CE products.	SSE has promoted cultural meaningfulness of areas, buildings and practices. Integration of these aspects would enhance the socio-cultural dimensions of the CE.
Education	Madrid: Elements of SSE are incorporated into the formal basic and secondary education in Madrid.	Education is needed to advance CE models (Andrews 2015; Kirchherr and Piscicelli 2019).
Resilience and preparedness for risks	Montevideo: Social resilience is promoted; citizens should be aware and perceptive of risks.	SSE can enrich CE to integrate resilience aspects to become "future proof" (Aguiñaga et al. 2018).

Source: compiled by authors.

governance. Regarding this challenge, future research would benefit from new perspectives and experimental orientation to understand what these suitable circumstances entail. For instance, the relational approach to urban governance (e.g., Bartels 2020) provides possibilities to think differently and analyse whether SSE and CE create circumstances by themselves and thereby entail novel opportunity streams for social sustainability.

Conclusion

Whereas CE has a strong appeal for the economic and waste management dimension of sustainability, the lack of consensus over the social dimension and its importance in CE constitutes a need to further untangle how and why sustainable cities should be both circular and solidary. This research identified a set of aspects based on the principles of SSE and studied three cases that can contribute to the social dimension of CE. Our exploratory study demonstrates that addressing the SSE principles in CE has the potential to enrich social sustainability in circular cities. In this way, SSE can contribute to addressing barriers to CE implementation, stakeholder interaction, social innovation, social inclusion, business models in sustainability transitions, cultural heritage, education, and resilience.

The results were promising. Firstly, the case of Hiedanranta showed that in suitable circumstances, spontaneous trajectories intrinsic to urban socio-cultural development can emerge and manifest "solidarity aspects" in conjunction with CE, even in an unplanned manner. At the same time, the case of Montevideo illustrated that integrating SSE into policymaking and associating it with other plastic concepts, such as resilience, can water down the strengths of SSE's social aspects. We also revealed that using SSE to enrich CE creates some practical and epistemological traps. The boundaries between SSE and CE are sensitive due to their ideological differences, and we are only beginning to understand the potential spaces where SSE and CE can interact and overlap in relevant ways for further research. The presented narratives regarding alternative economies come from different scholarly and practice traditions as well as diverse areas of the world. While SSE is associated with a critique of capitalism, CE seems to approximate green capitalism paradigms, which may present non-compatible worldviews.

In this regard, our exploratory findings are not of a universal nature. Rather, they refer to the contexts of these three cities, which bring interesting dimensions that can be further explored by scholarly research in the future. For example, which aspects and features of these meta-concepts are emphasised in their implementation to urban policies? What type of targets have been placed on the SSE and the CE informed policies, and how are these policies assessed and measured? When SSE and CE are jointly implemented on regional and national levels, how is the diversity of ideas which these concepts entail being accommodated? And, in the same context, how do actors select which aspects of SSE will go through "the final cut" of implementation, like in Saguier and Brent's (2017) case, where SSE was lessened as employment objectives?

Furthermore, for future research, more primary data should be collected from circular and solidarity cities in order to better inform the contradictions and complementarities of the two approaches. In conclusion, case studies that focus on specific cultural and social contexts are needed to achieve a general understanding of the synergy between circularity and solidarity.

References

Aguiñaga, Eduardo, Irene Henriques, Carlos Scheel, and Andrea Scheel. "Building resilience: A self-sustainable community approach to the triple bottom line." *Journal of Cleaner Production* 173 (2018): 186–196.

Alatalo, Elina, Jenni Kuoppa, Mikko Kyrönviita, and Markus Laine. "From ideas competition to citizens' vision: Planning Hiedanranta in follow-on workshops." University of Tampere & City of Tampere. 2017. https://www.e-julkaisu.fi/tampereen _kaupunki/from-ideas-competition-to-citizens-visions/mobile.html#pid=1

Andrews, Deborah. "The circular economy, design thinking and education for sustainability." *Local Economy* 30, no. 3 (2015): 305–315.

Angotti, Thomas. "Latin American urbanization and planning: Inequality and unsustainability in North and South." *Latin American Perspectives* 23, no. 4 (1996): 12–34.

Bartels, Koen. "Transforming the relational dynamics of urban governance: How social innovation research can create a trajectory for learning and change." *Urban Studies*, 57, no. 14 (2020): 2868–2884.

Bartone, Carl. "Environmental challenge in third world cities." *Journal of the American Planning Association* 57, no. 4 (1991): 411–415.

Béland, Daniel. *How ideas and institutions shape the politics of public policy.* Cambridge: Cambridge University Press, 2019. https://doi.org/10.1017/9781108634700

Brand, Ulrich, and Markus Wissen. "Strategies of a green economy, contours of a green capitalism." In *Handbook of the International Political Economy of Production*, pp. 508–523. Edward Elgar Publishing, 2015.

Brightman, Marc, and Jerome Lewis. "Introduction: The anthropology of sustainability: Beyond development and progress." In *The Anthropology of Sustainability*, 1–34. New York: Palgrave Macmillan, 2017.

Brown, Arthur Joseph. *Introduction to the world economy*, vol. 2. Oxon: Routledge, 2013. First published in 1959.

Bulkeley, Harriet, Lars Coenen, Niki Frantzeskaki, Christian Hartmann, Annica Kronsell, Lindsay Mai, Simon Marvin, Kes McCormick, Frank van Steenbergen, and Yuliya Voytenko Palgan. "Urban living labs: Governing urban sustainability transitions." *Current Opinion in Environmental Sustainability* 22 (2016): 13–17.

Calvo, Sara, and Andrés Morales. Introduction: Unravelling the global terms of the social and solidarity economy. In *Social and solidarity economy: The world's economy with a social face*, edited by Calvo and Morales, pp. 1–7. New York. Routledge, 2007

Cárdenas, Alzate, Martha del Socorro, Olga Lucía Arboleda Álvarez, and Oswaldo Antonio Salgado Cañaveral. "Retos de la Alcaldía de Medellín para la aplicación de la Política Pública de Economía Social y Solidaria." *Revista Finanzas y Política Económica* 8, no. 2 (2016): 401–430.

Corvellec, Hervé, Steffen Böhm, Alison Stowell, and Francisco Valenzuela. "Introduction to the special issue on the contested realities of the circular economy." *Culture and Organization*, 26, no. 2, (2020): 97–102.

Creswell, John W. *Qualitative inquiry and research design: Choosing among five approaches*. London: Sage publications, 2013.

Cruz, Antonio. "A construção do conceito de economia solidária no Cone Sul." *Revista de Estudios Cooperativos* 12, no. 1 (2007).

Dacheux, Eric, and Daniel Goujon. "The solidarity economy: An alternative development strategy? " *International Social Science Journal* 62, no. 203–204 (2011): 205–215.

Eisenhardt, Kathleen M. "Building theories from case study research." *Academy of Management Review* 14, no. 4 (1989): 532–550.

Esber Elias, Gabriel. "Solidarity Economy in the context of the Nordic welfare state." Master's thesis. Faculty of Management, Tampere University, Finland, 2019.

Fistola, Romano. "The unsustainable city. Urban entropy and social capital: The needing of a new urban planning." *Procedia Engineering* 21 (2011): 976–984.

Fitch-Roy, Oscar, David Benson, and David Monciardini. "Going around in circles? Conceptual recycling, patching and policy layering in the EU circular economy package." *Environmental Politics* 29, no. 6 (2020): 983–1003.

Foster, John Bellamy. "The absolute general law of environmental degradation under capitalism." *Capitalism, Nature, Socialism*, *3*(3) (1992): 77–81.

Fratini, Chiara Farné, Susse Georg, and Michael Søgaard Jørgensen. "Exploring circular economy imaginaries in European cities: A research agenda for the governance of urban sustainability transitions." *Journal of Cleaner Production* 228 (2019): 974–989.

Gallie, Walter Bryce. "Essentially contested concepts." In Proceedings of the Aristotelian society, vol. 56, pp. 167–198. Aristotelian Society, Wiley, 1956.

Geissdoerfer, Martin, Paulo Savaget, Nancy MP Bocken, and Erik Jan Hultink. "The circular economy: A new sustainability paradigm? " *Journal of Cleaner Production* 143 (2017): 757–768.

Geissdoerfer, Martin, Doroteya Vladimirova, and Steve Evans. "Sustainable business model innovation: A review." *Journal of Cleaner Production* 198 (2018): 401–416.

Ghisellini, Patrizia, Catia Cialani, and Sergio Ulgiati. "A review on circular economy: The expected transition to a balanced interplay of environmental and economic systems." *Journal of Cleaner production* 114 (2016): 11–32.

Gregson, Nicky, Mike Crang, Sara Fuller, and Helen Holmes. "Interrogating the circular economy: The moral economy of resource recovery in the EU." *Economy and Society* 44, no. 2 (2015): 218–243.

Gutberlet, Jutta, Sebastián Carenzo, Jaan-Henrik Kain, and Adalberto Mantovani Martiniano de Azevedo. "Waste picker organizations and their contribution to the circular economy: Two case studies from a global south perspective." *Resources* 6, no. 4 (2017): 52.

Haas, Willi, Fridolin Krausmann, Dominik Wiedenhofer, and Markus Heinz. "How circular is the global economy? An assessment of material flows, waste production, and recycling in the European Union and the world in 2005." *Journal of Industrial Ecology* 19, no. 5 (2015): 765–777.

Hahnel, R. (2014). *The ABCs of political economy: A modern approach. Hahnel, Robin. The ABCs of political economy: A modern approach.* London: Pluto Press, 2014. https://doi.org/10.2307/j.ctt183p3sv

Harris, Jerry. "Can green capitalism build a sustainable society? " *International Critical Thought* 3, no. 4 (2013): 468–479.

Hobson, Kersty. "'Small stories of closing loops': Social circularity and the everyday circular economy." *Climatic Change* 163, no. 1 (2020): 99–116.

Hobson, Kersty, and Nicholas Lynch. "Diversifying and de-growing the circular economy: Radical social transformation in a resource-scarce world." *Futures* 82 (2016): 15–25.

Kębłowski, Wojciech, Deborah Lambert, and David Bassens. "Circular economy and the city: An urban political economy agenda." *Culture and Organization* 26, no. 2 (2020): 142–158.

Kenis, Anneleen, and Matthias Lievens. "Searching for 'the political' in environmental politics." *Environmental Politics* 23, no. 4 (2014): 531–548.

Kingdon, John W. *Agendas, alternatives, and public policies*. New York: Harper Collins, 1995.

Kirchherr, Julian, and Laura Piscicelli. "Towards an education for the circular economy (ECE): Five teaching principles and a case study." *Resources, Conservation and Recycling* 150 (2019): 104406.

Kirchherr, Julian, Denise Reike, and Marko Hekkert. "Conceptualizing the circular economy: An analysis of 114 definitions." *Resources, Conservation and Recycling* 127 (2017): 221–232.

Korhonen, Jouni, Cali Nuur, Andreas Feldmann, and Seyoum Eshetu Birkie. "Circular economy as an essentially contested concept." *Journal of Cleaner Production* 175 (2018): 544–552.

Langhelle, Oluf. "Sustainable development: Linking environment and development." *Conceptual Innovation in Environmental Policy* edited by Meadowcroft, James, and Daniel, Fiorino, MIT Press (2017): 181–206.

Laville, Jean-Louis. "A economia solidária: um movimento internacional." *Revista crítica de ciências sociais* 84 (2009): 7–47.

Lazarevic, David, and Helena Valve. "Narrating expectations for the circular economy: Towards a common and contested European transition." *Energy Research & Social Science* 31 (2017): 60–69.

Lee, S. "Role of social and solidarity economy in localizing the sustainable development goals." *International Journal of Sustainable Development & World Ecology* 27, no. 1 (2019): 65–71.

Leino, Helena, and Eeva Puumala. "What can co-creation do for the citizens? Applying co-creation for the promotion of participation in cities." *Environment and Planning C: Politics and Space* 39 ,no. 4 (2021): 781–799.

Loiseau, Eleonore, Laura Saikku, Riina Antikainen, Nils Droste, Bernd Hansjürgens, Kati Pitkänen, Pekka Leskinen, Peter Kuikman, and Marianne Thomsen. "Green economy and related concepts: An overview." *Journal of Cleaner Production* 139 (2016): 361–371.

McDowall, Will, Yong Geng, Beijia Huang, Eva Barteková, Raimund Bleischwitz, Serdar Türkeli, René Kemp, and Teresa Doménech. "Circular economy policies in China and Europe." *Journal of Industrial Ecology* 21, no. 3 (2017): 651–661.

Meadowcroft, James, and Daniel J. Fiorino, eds. *Conceptual innovation in environmental policy*. Cambridge, MA: MIT Press, 2017.

Medina-García, Clara, Rosa de la Fuente, and Pieter Van den Broeck. "Exploring the emergence of innovative multi-actor collaborations toward a progressive urban regime in Madrid (2015–2019)." *Sustainability*, 13(1) (2021): 415.

Miller, Ethan. "Solidarity economy: Key concepts and issues." *Solidarity economy I: Building alternatives for people and planet*, 25–41, 2010.

Mochizuki, Yoko, and Masaru Yarime. "Education for sustainable development and sustainability science". In *Routledge handbook of higher education for sustainable development*, 11–24, 2016.

Morandeira-Arca, Jon, Enekoitz Etxezarreta-Etxarri, Olatz Azurza-Zubizarreta, and Julen Izagirre-Olaizola. "Social innovation for a new energy model, from theory to action: Contributions from the social and solidarity economy in the Basque Country." *Innovation: The European Journal of Social Science Research* (2021): 1–27.

Moreau, Vincent, Marlyne Sahakian, Pascal Van Griethuysen, and François Vuille. "Coming full circle: Why social and institutional dimensions matter for the circular economy." *Journal of Industrial Ecology* 21, no. 3 (2017): 497–506.

Moyer, Jonathan D., and Steve Hedden. "Are we on the right path to achieve the sustainable development goals? " *World Development* 127 (2020): 104749.

Murray, A., Skene, K. & Haynes, K. The circular economy: An interdisciplinary exploration of the concept and application in a global context. *Journal of Business Ethics* 140, 369–380 (2017).

Nilsson, David. "A heritage of unsustainability? Reviewing the origin of the large-scale water and sanitation system in Kampala, Uganda." *Environment and Urbanization* 18, no. 2 (2006): 369–385.

North, Peter, and Molly Scott Cato, eds. *Towards just and sustainable economies: The social and solidarity economy north and south.* Bristol: Policy Press, 2017.

Nylén, Erkki-Jussi. "Kiertotaloussiirtymä ja uudet markkinat." *Alue ja ympäristö* 48, no. 1 (2019): 14–28.

Nylén, Erkki-Jussi., Riki, Aleksandra., Jokinen, Ari. and Jokinen, Pekka. "Kiertotalouden kestävyyslupaukset. Espoon, Lahden, Tampereen ja Turun kaupunkistrategioiden vertailu." *Yhteiskuntapolitiikka* 86, no. 4 (2021).

Padilla-Rivera, Alejandro, Sara Russo-Garrido, and Nicolas Merveille. "Addressing the social aspects of a circular economy: A systematic literature review." *Sustainability* 12, no. 19 (2020): 7912.

Poirier, Yvon. "Social solidarity economy and related concepts origins and definitions: An international perspective." *Socioeco. org.* 2014. http://www. socioeco. org/bdf_fiche-document-3293_en. html.

Pollitt, Christopher, and Peter Hupe. "Talking about government: The role of magic concepts." *Public Management Review* 13, no. 5 (2011): 641–658.

Prendeville, Sharon, Emma Cherim, and Nancy Bocken. "Circular cities: Mapping six cities in transition." *Environmental Innovation and Societal Transitions* 26 (2018): 171–194.

Quiroz-Niño, Catalina, and María Ángeles Murga-Menoyo. "Social and solidarity economy, sustainable development goals, and community development: The mission of adult education & training." *Sustainability* 9, no. 12 (2017): 2164.

Razeto, Luis. "La economía de solidaridad: concepto, realidad y proyecto." *Persona y sociedad* 13, no. 2 (1999): 15.

Raworth, Kate. *Doughnut economics: seven ways to think like a 21st-century economist.* Vermont. Chelsea Green Publishing, 2017.

Rees, William E. "Is 'sustainable city' an oxymoron?" *Local Environment* 2, no. 3 (1997): 303–310.

Romano, Bernardino, and Francesco Zullo. "The urban transformation of Italy's Adriatic coastal strip: Fifty years of unsustainability." *Land Use Policy* 38 (2014): 26–36.

Sabedot, Sydney, and Tiago José Pereira Neto. "Desempenho ambiental dos catadores de materiais recicláveis em Esteio (RS)." *Engenharia Sanitária e Ambiental* 22, no. 1 (2017): 103–109.

Sachs, Jeffrey D., Guido Schmidt-Traub, Christian Kroll, Guillaume Lafortune, Grayson Fuller, and Finn Woelm. *Sustainable Development Report 2020.* 2020.

Saguier, Marcelo, and Zoe Brent. "Social and solidarity economy in South American regional governance." *Global Social Policy* 17, no. 3 (2017): 259–278.

Särkilahti, Maarit, Maria Åkerman, Ari Jokinen, and Jukka Rintala. "Temporal challenges of building a circular city district through living-lab experiments." *European Planning Studies* (2021). DOI: 10.1080/09654313.2021.1965963

Savini, Federico. "The economy that runs on waste: Accumulation in the circular city." *Journal of Environmental Policy & Planning* 21, no. 6 (2019): 675–691.

Schöggl, Josef-Peter, Lukas Stumpf, and Rupert J. Baumgartner. "The narrative of sustainability and circular economy-A longitudinal review of two decades of research." *Resources, Conservation and Recycling* 163 (2020): 105073.

Singer, Paul. "Economia solidária: um modo de produção e distribuição." *A economia solidária no Brasil: A autogestão como resposta ao desemprego*, pp. 11–28. São Paulo: Contexto, 2000.

Sullivan, Sian. "Green capitalism, and the cultural poverty of constructing nature as service-provider." *Radical Anthropology* 3 (2009): 18–27.

Thiele, Leslie Paul. "Human rights at the 'end of nature'." *Journal of Human Rights* 18, no. 1 (2019): 19–35.

Tienhaara, Kyla. "Varieties of green capitalism: Economy and environment in the wake of the global financial crisis." *Environmental Politics* 23, no. 2 (2014): 187–204.

Turku, Veera, Ari Jokinen, and Pekka Jokinen. "Elävät kaupunkilaboratoriot: kulttuuritoimijat Tampereen Hiedanrannan uudistajina." *Hallinnon Tutkimus* 40, no. 1 (2021): 53–68.

United Nations. "Transforming our world: The 2030 Agenda for Sustainable Development". (2015). Accessed January 2021. https://sdgs.un.org/2030agenda

United Nations. "The future is now–science for achieving sustainable development." (2019). Accessed January 2021 https://reliefweb.int/sites/reliefweb.int/files/resources/24797GSDR_report_2019.pdf

Valenzuela, Francisco, and Steffen Böhm. "Against wasted politics: A critique of the circular economy." *Ephemera: Theory & Politics in Organization* 17, no. 1 (2017): 23–60.

Wardekker, Arjan, Bettina Wilk, Valerie Brown, Caroline Uittenbroek, Heleen Mees, Peter Driessen, Martin Wassen, Arnoud Molenaar, Jim Walda, and Hens Runhaar. "A diagnostic tool for supporting policymaking on urban resilience." *Cities* 101 (2020): 102691.

Weili, Liu, and Hayat Khan. "A literature review on the definition of sharing economy." *Global Economy Journal* 20, no. 3 (2020): 2030001.

Williams, Joanna. "Circular cities." *Urban Studies* 56, no. 13 (2019): 2746–2762.

Winans, Kiara, Alissa Kendall, and Hui Deng. "The history and current applications of the circular economy concept." *Renewable and Sustainable Energy Reviews* 68 (2017): 825–833.

World Bank. "Urban development". Accessed August 2020. https://data.worldbank.org/topic/urban-development

4 Towards a socially inclusive circular economy

A study of tenant engagement in European social housing organisations

Halima Sacranie and Sultan Çetin

Introduction

Cities worldwide face significant socio-economic and climate change-related challenges, which have become even more evident with the impacts of the coronavirus pandemic. Growing urban populations, inadequate housing supply, and other context-specific problems reduce accessibility to affordable housing, putting millions of lower-income households at risk. According to the recent *State of Housing in Europe* report, 1.1 million affordable homes are needed in England alone, and around 700,000 people are homeless, sleeping rough, or in temporary locations across the European Union (EU) (Housing Europe 2021). As a response, most European countries have set objectives to build new affordable homes, e.g., France is investing in around 110,000 new social housing units (Housing Europe 2021). However, a critical question remains of how to construct these much-needed homes without damaging the natural environment.

The construction sector typically operates in a linear production-consumption system and has an enormous impact on the environment. The system inputs are significant in volume as the industry uses around 50% of raw materials (Herczeg et al. 2014) and consumes around 40% of total energy (European Commission 2020) for the production and operation of buildings in Europe. Similarly, waste and greenhouse gas emissions are created as system outputs, which account for 35.9% and 36% of all sectors respectively (European Commission 2020; Eurostat 2018). Considering the rising concerns over climate change and the construction sector's contribution to it, a radical change is needed in how buildings are produced, function, and are used.

An alternative model to address some of these challenges is the circular economy (CE). As opposed to linear production and consumption systems, CE focuses on minimising resource inputs and outputs by creating cyclic flows to narrow (resource efficiency), slow (lifetime extension), and close (recycle waste) resource loops (Bocken et al. 2016). CE derives conceptually from existing models such as industrial ecology, cradle-to-cradle, and the performance economy. Despite numerous attempts, there is no widely accepted definition of CE. Perhaps the most popular CE definition is the one proposed by the Ellen MacArthur Foundation (EMF) (Kirchherr, Reike, and Hekkert 2017), which interprets CE as

DOI: 10.4324/9781003255246-4

an industrial system that is "restorative or regenerative and aims to shift towards using renewable resources and eliminating toxic chemicals and waste through superior design and business models" (EMF 2012). The EMF has played a critical role in defining and conceptualising CE to decouple economic growth from raw material consumption (EMF 2012) and offers technical solutions to achieve this grand objective.

As an emerging research topic, CE in the construction industry has predominantly investigated economic and environmental benefits in building production processes. Researchers concentrate on recycling or reusing construction and demolition waste, reducing raw material use, and eliminating waste throughout the lifecycle stages of buildings (Munaro, Tavares, and Bragança 2020). Such technical approaches focus on the production and end-of-life phases (e.g., design for disassembly) and overlook social aspects such as well-being, equity, and behaviour change (Murray, Skene, and Haynes 2015). Indeed, the literature purports that the social dimension of CE is a neglected area, apart from a limited focus on mainly labour markets, health and safety, and participation approaches (Padilla-Rivera, Russo-Garrido, and Merveille 2020). Geissdoerfer et al. (2017) also argue that CE offers clear advantages mainly for the economic actors, benefiting the economy and environment while neglecting explicit benefits for society. Merli, Preziosi, and Acampora (2018) reiterate that CE research prioritises environmental sustainability but only marginally considers the social aspect of the triple bottom line that should constitute the foundation of this sustainable approach.

Other scholars contend that the social objective of CE is the sharing economy where a cultural shift is expected from being a consumer towards being a cooperative and community user of the physical material capacity (Korhonen, Honkasalo, and Seppälä 2018). Furthermore, Moreau et al. (2017) stress the essential role of nontechnical, institutional, and social aspects in developing CE and further propose the social and solidarity economy (SSE) as a means to fill the elusive social gap of the CE concept. According to the authors, in an SSE, "societal decisions could be made in terms of what materials should be reduced or reused, or what materials should be recycled as a priority, toward a common good, regardless of economic profitability" (503). Such a democratised participative approach appears not to be a priority in the current implementation of CE in the construction sector.

In this chapter, we explore the neglected social dimension of CE through case studies of four social housing organisations in Europe, which have undertaken new circular building and renovation projects. The focus of our investigation is on tenant engagement within these circular social housing projects, building on the arguments of Moreau et al. (2017) and considering how models of tenant engagement link to principles around participation, collaboration, and citizenship.

An interesting dimension to this research is that these case study social housing organisations (SHOs)[1] are, by definition, hybrid or third-sector organisations with a not-for-profit business model, providing decent, affordable housing for economically and socially disadvantaged groups in society. They manage or own significant European housing stock and are responsible for housing production, regeneration, and maintaining a good quality of homes and neighbourhoods. These organisations, therefore, have an implicit social purpose.

CE principles offer several promising opportunities for SHOs, including improving the environmental performance of housing production; decreasing dependency on raw materials; and therefore strengthening SHOs' financial security against rising material prices, allowing them to build more social homes with the savings or operating surpluses resulting from the circular management of resources. Thus, there is a strong argument that implementing CE also has a de facto social impact through any economic or performance improvement for these social-purpose housing organisations.

However, beyond these inherent social impacts through improved organisational performance and efficiencies, our research seeks to understand the more direct social outcomes of circular housing for the people who occupy these homes. We explore, therefore, changes in the interactions or engagement between the SHOs and their tenants as these organisations implementing CE strategies take into consideration their tenants' perspectives due to the SHOs' social responsibilities (Eikelenboom, Long, and de Jong 2021). Through our case studies, we investigate if and how SHOs capture evidence on the social dimensions of CE as pertaining to their tenants, and how they engage their tenants in these circular projects; whether these processes remain as "business as usual" or if a "new circular engagement" emerges.

The following sections of this chapter set out an overview of the current state of CE implementation in the social housing sector in Europe, explaining our case for investigating circular social housing and our focus on tenant engagement linked to SSE to consider a collaborative and socially inclusive CE. The research methodology for this study is outlined and a brief profile of the case study organisations is provided before the case study findings are discussed and outcomes mapped theoretically. The concluding section of this paper reflects on the evidence presented through the case studies on "business as usual" versus a "new circular engagement" and sets out the next steps for this ongoing research.

Towards a collaborative and socially inclusive circular economy in social housing

In the EU, around 76% of the overall building stock comprises residential buildings (European Commission 2013), of which roughly 9% is managed or owned by SHOs (Housing Europe 2021). The size, definition, and responsibilities of the SHOs vary across Europe. The Netherlands, Austria, and Denmark have the most extensive social housing stock of 29.1%, 24%, and 21%, respectively (Housing Europe 2021). In contrast, other EU countries like Greece do not have a social housing system. Furthermore, there is no universal definition of social housing as ownership, tenure, and allocation, and national policies vary across countries. Social housing can generally be considered a system that provides long-term housing with below-market rents to households with limited financial resources, which is usually supported by public or private subsidies (Granath Hansson and

Lundgren 2018). Public actors or non-profit SHOs play a crucial role in housing provision. For example, in the Netherlands, SHOs are "independent from the state and act as a non-profit enterprise that pursues social goals within a strict framework of national laws and regulations by involving local government, tenants and other stakeholders in their policies and are accountable to society" (AEDES 2016: 3).

CE is a new phenomenon in the social housing sector as energy transition and decarbonisation of the housing stock has been the focal point of attention for environmental sustainability. Following the EU's first CE action plan, the CE concept has had a growing interest within the construction sector, and several SHOs across the EU have started piloting circular construction projects. According to Çetin, Gruis, and Straub, the first experimentations of SHOs concern the development of technical solutions in housing production for resource efficiency, waste reduction and future re-use of building materials rather than social dimensions or cultural behaviour change of residents (Çetin, Gruis, and Straub 2021). The authors further explain that for Dutch housing providers, "lack of tenant awareness and interest" towards CE is not a barrier when implementing CE principles in pilot projects as tenants are usually not involved in the project development phase. However, Eikelenboom, Long, and de Jong explore how social elements could be integrated into CE strategies of SHOs, establishing relationships with communities in the vision creation, and involving them in the execution of CE strategies, thereby leveraging community support for the environmental targets of these strategies (Eikelenboom, Long, and de Jong 2021).

To build a conceptual understanding of collaborative and socially inclusive circular social housing, this section triangulates literature in the areas of social housing tenant engagement and SSE, linking tenant engagement theories with models adopted in CE literature around social inclusion, citizenship, and the social and solidarity economy (SSE).

In the context of this study, we explore tenant engagement in the design and planning stages of their housing providers' CE construction and renovation projects. More broadly though, tenant engagement is critically important in social housing strategic management and operations through corporate governance, regulatory scrutiny, housing management services, consumer standards, and community investment practices.

A seminal theory regarded as the foundation for understanding tenant engagement in social housing is Arnstein (1969)'s ladder of participation, which locates varying degrees of participation along the rungs of a ladder model starting from non-participation such as manipulation, progressing to tokenistic participation (like informing and consultation), before reaching the higher echelons of citizen power which include partnership, delegated power, and finally citizen control. These ideas of citizenship and collaborative decision-making have clear parallels to the principles of SSE with regard to social inclusion and empowerment.

Apart from the aspirational and intrinsic value of citizenship and participation, Hickman and Preece (2019) describe the practical benefits of tenant participation through commercial benefits to the housing service, individual benefits to tenants and social benefits to local communities (Hickman and Preece 2019). Tenant engagement may be collective or individual, the latter usually relating more to a customer or consumer approach. There are also overlaps between tenant engagement and community investment (CI), where SHOs have priorities around facilities, projects, and schemes for their tenants such as community centres, nurseries, or employment and skills training. Tenant engagement may feed into SHO community investment strategies, although shifts from community-based to customer-focused models may occur due to financial and efficiency drivers and changing organisational cultures (Sacranie 2012). Indeed, organisational culture is regarded as a key aspect affecting the type of tenant engagement adopted by housing organisations (Pawson et al. 2012). (This would be an interesting concept to explore through the growing adoption of CE in social housing organisations, considering organisational cultural shifts and paradigms which might emerge in this new field of circular social housing.)

An often-cited framework on tenant engagement is Cairncross, Clapham, and Goodlad's (1994) *consumerist, citizenship*, and *traditional* model. Consumerist models of tenant engagement see engagement as a customer service or satisfaction improvement tool with commercial benefits around cost efficiency and asset management through tenant scrutiny panels and satisfaction surveys. Citizenship models are about meaningful and inclusive engagement through shared decision-making, collaboration, and empowering tenants (as captured customers with limited alternative housing options) to make decisions about their housing service and are seen as intrinsically valuable through the actual process of engagement beyond what the engagement itself achieves (Hickman and Preece 2019). Benefits of this approach include embedding social interaction and social cohesion in communities and improving relationships between tenants and landlords. Finally, traditional tenant engagement is typically driven by SHOs in a clear top-down approach, with the housing provider "looking after" or guarding the best interests of their tenants. Different approaches are not mutually exclusive and can co-exist within the same organisation.

Mullins, Shanks, and Sacranie (2017) reiterate that a key debate around the identity of tenants and their role in housing governance is that of consumerist versus citizenship approaches to tenant engagement. Consumerist approaches frame tenants as consumers of a housing service and prioritise a customer service focus to engagement, often, but not universally, tied in with a commercial orientation of larger SHOs. Citizenship approaches are typically more participatory and prevalent in (often smaller) cooperatives and locally based housing organisations. However, as other cited authors suggest, the approaches can be blended in the manner of adoption.

As outlined in this chapter introduction, a key research aim for this study is to move beyond a de facto social purpose to understand approaches by which the social aspect of CE in social housing may be framed, such as social innovation, the sharing economy, and SSE, which hold common principles around community, collaboration, and democratised participation. These fundamental principles are also characteristic of the citizenship approaches to tenant engagement in social housing (Cairncross, Clapham, and Goodlad 1994) thereby supporting the investigation of tenant engagement practices in these case study housing projects as a way to understand the social dimension of CE in social housing.

Having a shared understanding, values, sense of ownership and purpose are regarded as integral to social cohesion (Berger-Schmitt 2002; Forrest and Kearns 2016). Social housing communities are connected as networks through shared community investment and social value projects but also by tenants who are involved in decision-making around the housing and neighbourhood management services. The social innovation approach (Marchesi and Tweed 2021) encompasses grassroots engagement initiatives through community interest groups and citizen engagement which help embed sustainable behavioural changes and practices. Developing a typology of seven community social innovation (SI) initiatives ranging from community growing to social enterprises, Marchesi and Tweed (2021) argue through case study analysis that SI initiatives can benefit both communities and cities by creating "alternative production-consumption practices into their communities to reduce waste, save money, create a more cohesive community, enhance people's skills, and increase community ownership" (Marchesi and Tweed 2021: 12). The authors, therefore, make a case for SI approaches to be integrated into CE models in order that SHOs could benefit from including their tenants in their transitions to CE.

Woodard and Rossouw's (2021) study on engaging social housing residents implementing a circular economic waste management programme found that participants identified embedding "a sense of community" as one of the main positive outcomes for the project and that a key motivation for their involvement was to increase community spirit and social cohesion. Environmental improvements were noted, but additionally, the collaborative approach employed engendered a sense of pride in the local neighbourhoods, improved well-being, and fostered better relationships with local authority staff and contractors. Residents also found the interactive and inclusive engagement made them feel valued as stakeholders in an important project and that the engagement was meaningful rather than tokenistic because of this bottom-up approach.

Another complementary model reiterating these principles of collaboration and social inclusion is the sharing economy, which has also been considered in relation to the social and cultural vacuum of CE models of sustainability. Through a sharing economy, cooperative community rather than consumerist customer culture can enhance the social impact of CE:

> The social objective is the sharing economy, increased employment, participative democratic decision-making and more efficient use of the existing physical-material capacity through a cooperative and community user (user groups using the value, service and function) as opposed to a consumer (individuals consuming physical products) culture.
>
> (Korhonen, Honkasalo, and Seppälä 2018)

Reinforcing these common principles applied to CE, Moreau et al. (2017) argue that SSE is the ideal model to address the missing social dimensions of CE by "bringing together cooperative efforts and promoting democratic participation in economic activities," which can lead to "more robust CE strategies toward societal and environmental aims". SSE aims to create social value through this participative decision-making towards a democratised economy (Moreau et al. 2017). Indeed, "the models of collaborative and democratic governance systems, which challenge the profit motive" are already evident in third sector housing organisations, which are not-for-profit, driven by social purpose and already integrate some form of tenant engagement as part of their governance or regulatory obligations.

Four circular social housing case studies

This study is based on longitudinal multiple-case study research (Yin 2018) with four circular social housing organisations in the Netherlands, France, Belgium, and the UK, respectively. The case projects and this research study are ongoing, and this chapter draws on early data collection through documentary evidence and interviews with the case SHOs.

Our research focused on the organisational engagement of tenants during the early project phase,[2] and how the role of tenants in the circular housing projects differed from the usual way in which these SHOs engaged with their tenants. We wished to explore specifically how the tenants of these four SHOs would be involved in the circular housing projects, from participation in planning and design to the allocation of tenants to these schemes. Would these interactions be incorporated into existing organisational processes and engagement models, or would the circular nature of the projects necessitate new processes and a different way of engaging?

The four case studies are part of an Interreg Northwest Europe funded project called CHARM (Circular Housing Asset Renovation and Management),[3] which aims to promote circular asset management approaches in the social housing sector to prevent downcycling of materials in renovation and construction of social rented dwellings. By developing and implementing demonstration circular exemplars, the case study SHOs hope to recover 40,000 tonnes of materials annually. Table 4.1 provides an overview of the four case study organisations.

Case 1 is a housing and care provider operating in the UK and serves around 54,000 residents. The environment policy of Case 1 includes saving energy, reducing waste, protecting wildlife and biodiversity, preventing pollution, and cutting the amount of carbon emissions generated by their operations, staff, and tenants. They use waste hierarchy in their daily operations and see CE principles as the "reuse" element. Within the context of CHARM, Case 1 is developing 12 virtually plastic-free homes. Case 2 is one of France's largest social housing providers, serving around 287,000 residents, and allocates 3,750 housing units each year to low-income citizens. This SHO is committed to social and ecological sustainability and supports the vision of a carbon-neutral city alongside the Paris Climate agreement. Case 2 aims to increase reusing of materials in their existing building stock, testing CE principles with several light and deep renovation projects within the CHARM project.

Table 4.1 Circular social housing case study profiles

Case SHO and location	Number of homes/residents (approximately)	CE aspiration	CE case study projects/ potential number of tenants impacted
Case 1, The UK	25,000/54,000	CE is the 'reuse' element in waste hierarchy	12 virtually plastic-free social rented homes; 12 tenants
Case 2, France	125,700/287,000	CE is seen as an opportunity to increase material reuse in renovation and newly built housing projects	3 light building renovations; 2 deep building renovations; and 2 transformation projects/over 1,000 tenants
Case 3, The Netherlands	32,900/70,000	CE is a primary sustainability goal in roadmap to be fully circular in ten years	18 new housing units by reusing materials from other demolition sites; 18 to 28 tenants
Case 4, Belgium	2,350/4,000	CE as an underlying ethos with a six-step operational CE approach	40 deep renovations and two new apartments; 90 tenants

Case 3 is a Dutch SHO serving around 70,000 tenants. They are one of the forerunners in implementing CE in the country with strong commitments towards circularity through their strategic planning and operations. Within the scope of CHARM, Case 3 is developing 18 new housing units. Finally, Case 4 is a relatively small SHO operating around 2,350 homes in Belgium and perceives CE as a philosophy rather than a new approach in housing production. The company applies a six-step CE approach in their operations: decrease demand for natural resources; use renewable resources; use resources efficiently; measure, monitor, and evaluate; coach users; share information and outputs with all stakeholders. Their CHARM projects include the renovation of 40 dwellings with reused materials and building two new apartments reusing materials from demolitions sites.

Case study findings

Case 1

For their CE project, Case 1 were developing 12 plastic-free homes as part of a broader town centre regeneration programme. This scheme was also part of a cooperative subsidiary of the SHO, which meant the cooperative structure with existing tenant engagement mechanisms would be employed, however, with some exceptions. Due to the nature of the site and tight planning restrictions around the town centre regeneration site, the development was restricted to apartments only, while project costs per unit meant the size of these 12 apartments was limited to 35 sq m. Furthermore, the innovative replacement of plastic materials for sustainable alternatives meant limited early development design choices. For this circular project, a pre-development tenant design panel was not established as was customary for other developments. The initial proposal for this scheme had been presented to the cooperative tenant board, who were particularly interested in the fuel economy of these homes.

While some of the building technologies and material alternatives did introduce engineering complexities, for Case 1 the aim was not to sell or encourage a particular sustainable style of living but rather for the sustainable building to fit into the existing lifestyle of their tenants. The expectation was that internal specifications would be of a better quality than usual, but apart from that, the apartments would have no obvious differences, while any new technologies like the heating system would have to be explained to tenants.

> The house will function exactly the same as a normal house, and that's part of what we're intending to do. So, although it's been circular, there will be no noticeable difference to the end user, and that's the important part of our philosophy…We shouldn't be creating a lifestyle change. We should be making it easy for people to live their lifestyle in a green way. That's what we would say and that's what we think we've achieved here. We will obviously give people an explanation of the scheme and the ethos behind it. There will be a need to explain some technologies.
>
> (Project Manager, Case 1)

Tenants would be allocated to these new homes according to local council waiting lists based on general needs and suitability for the apartments. This was expected to be predominantly younger, single tenants or couples, who would not specifically be selected because of any predisposition towards living in a circular or sustainable home. Around six meetings were planned with tenants before they moved in to create a sense of community, awareness of the circular and sustainable characteristics of the scheme, and allow them to choose certain specification finishes on bathrooms, kitchens, flooring, and painting. Once tenants moved in, a check was planned to see how they settled in and managed their property.

> It's all of our normal tenant satisfaction surveys that we do for when people move into a property. We will almost certainly have a sort of debrief meeting with residents to work out what's going well and what's gone badly after sort of 6 or 12 months.
>
> But we won't be doing anything special around this development that we wouldn't do on any other development.
>
> (Project Manager, Case 1)

Although the social impacts of these circular homes were not intended to be formally evaluated, the project manager for the circular homes anticipated that there would be additional benefits for tenants which would include lower heating costs; cleaner air due to the timber-framed construction, reduced plastics and plaster, and better noise insulation (which would reduce the impact of anti-social behaviour).

Overall, an interesting aspect of Case Study 1 was that traditional co-design approaches with tenant design panels were restricted because of planning parameters for the building site and the technical specifications for the circular dwellings. While there is added complexity in the engineering of heating and plastic alternatives, the aim for these circular homes is to be user-friendly and straightforward for tenants. Tenant engagement will be "business as usual" once the houses are allocated, apart from initially explaining to tenants the new sustainable features of their homes, and the choice of certain finishes to personalise their homes.

Case 2

Case 2's circular project involved the refurbishment and renovation of around 800 of their properties across different suburbs and estates where their housing stock is located and affected a range of tenants from different socio-economic and cultural backgrounds ranging from "extreme poverty" to more "middle-class" neighbourhoods. This SHO worked with architects across different sites to reuse, recycle, and upcycle parts, fixtures, and fittings such as re-commissioned timber doors or windows. Like any other major renovation works, approval was first sought by a tenant majority vote in the area/estate where the work was to be undertaken. Along with the vote, tenants also received individualised copies of any planned refurbishment of their home and building.

Tenant engagement typically occurs through structured public or town hall meetings and frontline housing management, drop-in clinics, and Case 2's outreach service. At these public meetings, members of local communities who are not necessarily Case 2's tenants will often attend as well. The organisation has found that tenants from middle-class backgrounds are more likely to engage meaningfully, whereas the more economically deprived and vulnerable tenants like refugees are less able to do so, while others may have language barriers not being first language French-speaking. Similarly, other groups, like single mums or others with dependent responsibilities who are already volunteering (for example, through school parents' associations), find it challenging to be involved in the co-design and development of a circular housing project.

Case 2 also has around ten voluntary tenant associations, but these do not cover all their stock areas. Architects work with tenants on a site-by-site basis to ensure tenants are satisfied with the renovations planned and understand and accept the choice of materials used. In some projects, the architects try to employ tenants to undertake some of the refurbishment work so that the tenants can learn employable skills, demonstrating a collaborative approach to both engagement and a key social impact of the SHO's circular project.

Community investment and inclusion are critical components of Case 2's outreach strategy. They work extensively with local partners and NGOs on the ground, considering the circular project as part of a broader socio-economic context which includes housing, education, and employment. While tenant acceptance was seen as crucial for the success of the circular project, the circular ethos driving these refurbishments did not seek to imbue tenants with a lifestyle or philosophy change, but rather to have the refurbishment work completed to a good standard.

> Why it's important to us (to use) reconditioned building materials is not a part of something that they see a lot...They don't know (about it), they know that we owe them work on their apartments and we do. And they know that we owe them a certain quality.
>
> (Project Coordinator, Case 2)

At the same time, a circular philosophy was being adopted more holistically organisationally with the support of Case 2's senior leadership. The hope was that this could be reflected across other housing management services and eventually begin to change social behaviours and enhance tenants' understanding of sustainable living. A visible example of this was having a large waste collection service for tenants by a recycling company:

> I think what is important for us is to have complimentary actions so that people can understand the concept of reuse more globally and how it can impact their everyday life...I think it (circular economy) is an organization-wide philosophy.
>
> (Project Manager, Case 2)

Case 3

Case 3's circular project involved building new circular homes rather than refurbishing existing properties. As a pioneering sustainable SHO adopting the principles of CE, they quickly identified a group of seven tenants (on short-term contracts) who were invited to take part in the pre-development process. It was hoped that these tenants would move into their new circular homes, and they began a process of weekly engagement and co-design on the new properties. The tenants were enthusiastic and engaged with the circular principles behind the development, becoming tenant ambassadors in organisational videos explaining the importance of the circular design of their new homes.

The design consultation included ergonomics and breaking component parts of the home down to electrical sockets and doorbells, questioning what was essential to include in a circular home and how those components might be reconfigured or redesigned to enhance sustainability. It was felt that this group of seven tenants had formed a little community who had become positively and intrinsically motivated around the circular principles and design aspects of their future homes.

However, due to planning and development delays on the project, these tenants engaged in the co-design process were offered and accepted alternative homes from Case 3. This reflected a critical challenge of inclusive and collaborative tenant engagement, which is most effective over a longer period of time, but which may become too long a process for potential tenants for whom finding a social home is a priority.

> It was not their goal to get involved in a project; it was a goal to get a home…
> In the end, they wanted a home and nothing more.
>
> (Project Manager, Case 3)

For the circular homes that would be developed, a new set of tenants would not be engaged in development co-design, but the learning from the initial co-creation workshops would be incorporated into any design modifications for the relocated building site. New tenants would be consulted to choose certain fixtures and fittings. Case 3 were still keen to follow through the lived-in experience of tenants on:

> How is it to live in a circular building, what works, what doesn't work.
>
> (Project Manager, Case 3)

Other tenant groups were typically engaged across their housing stock in different ways, and not always in co-design but sometimes around interesting projects like communal gardens. Internal layouts were more restricted but typically tenants could be involved in selecting kitchens and bathrooms. As part of regulatory and statutory requirements in the Netherlands, Case 3 also have a Tenant Advisory Board who play a role in the organisation's governance, agreeing on an annual corporate strategy for the following year. There are also tenant committees who are consulted on smaller projects, but these can be dominated by a few available members (three to five tenants representing an estate or area of 200 homes) and the development team, therefore, try to encourage a broader and more diverse engagement. Traditionally for renovation projects, Case 3 would do telephone surveys before the renovation, during, and six weeks after to see how the work impacts their tenants, but more recently, they had adopted a less intrusive approach.

> The thing is, we want to build for our regular tenants so not just for people who have a lot of involvement in sustainability.
>
> (Project Manager, Case 3)

For the circular projects, tenant engagement would be about knowledge transfer and educating tenants about the different features of their new home, like the alternative materials and heating systems, and any restrictions that might pertain to the different specifications. Inevitably because of the technical requirements for the circular homes, the material and finish choices would be more restricted.

It didn't seem necessary to highlight to tenants the recycled building materials for new circular schemes, as it would not be an issue to have used components in an older rental property. An example was given of the click dry bricks system that "looks and feels like a brick". Case 3 assumed the majority of their tenants would not care about this brick system as their priority was securing a comfortable and affordable home and not specifically a sustainable one. While the home would be innovative in terms of development, design, and construction, the tenant lived experience should be the same as usual.

Case 4

Case 4 had several concurrent circular projects, including the renovation of their offices ("leading by example") and an entire estate renovation. The organisation had embraced circularity as an organisational ethos, and all new developments and refurbishments would aim to be "as circular as possible". For Case 4, the pandemic had proved particularly challenging around engaging tenants. Their focus was rather on assisting vulnerable, lonely, and elderly tenants through this difficult period and less on prioritising interaction about their circular developments. Key stakeholders for Case 4 were the architects who were aligned to the same CE and sustainable design principles and values as the organisation. The focus of their circular project was around technical monitoring rather than tracking any social impact.

> We believe that the mindset of the architect is very important if you want to really get circularity in your design, in your building, in your innovation... And that is why we focus mainly now on the role of the architect, not as much as we want to focus on the tenants because they do not really have a choice... They get the dwelling, and it is what it is.
>
> (Project Manager, Case 4)

Tenant engagement for this housing association typically took the form of "Kitchen Table Talks" to gain insight into tenants' preferences regarding design and how involved they wished to be. However, the pandemic had meant the cessation of these in-person talks, and the organisation was considering alternative digital ways to communicate around their demonstration sites, including a video walkthrough. Community houses that operated on estates across the municipalities formed the usual hub for engagement with tenants, as social and community services were provided here, and tenants could raise issues around service delivery. Beyond that, there wasn't a formal governance structure of tenant boards and panels to consult, but as a small housing association with 25 staff members, frontline officers and care workers knew up to 90% of tenants on a first-name basis.

The renovation of existing properties using recycled materials meant tenants needed to be convinced and assured that they were still getting a "shiny and new" high-quality, sustainable product. At the time these case studies were conducted, Case 4 was in the planning stages for their circular renovation project and described the selection process for which homes would be renovated. Tenant willingness and agreement were essential, not only for the circular components but also because the renovation work would take place while the tenant still lived in the property. Tenant allocation to properties occurred in the same way as usual, based on locality, special needs, suitability for household size, and waiting lists. Different engagement processes were required with tenants for new developments ("it is very useful to get the tenants involved in an early stage") compared to tenants already living in existing properties chosen for refurbishment.

Tenant engagement for the circular projects was of particular importance to explain the sustainable strategy, and why, for example, when tenants expected a new kitchen, they would have a renovated one instead. A lack of construction skills and building insurance liabilities were seen as a barrier to involving tenants in the renovation work for their homes. As part of their sustainable ethos, however, Case 4 intended to set up shared tools schemes for tenants to maintain their properties, such as hedge trimmers, ladders, and tools to undertake minor repairs themselves and to give their tenants a sense of ownership of their homes.

Technical constraints on materials used meant less choice for tenants, and therefore engagement was more about informing and explaining. There was an acknowledgement that it would be challenging to change tenants' behaviour and attitude by housing them in circular dwellings. Nevertheless, Case 4 considered that a key part of their role was to demonstrate to tenants that if they changed their behaviour, it would increase the environmental impact of the building and they could also save on their energy bills. Trying to capture the social impact of a circular house on tenants' behaviour and attitudes would be difficult because the technical aspects and outputs of the circular projects around energy saving, material used, and carbon reduction impact measurements were already challenging to capture.

Mapping circular tenant engagement

A common rationale across all the case studies was around customer satisfaction and providing a quality affordable home for their tenants that incidentally happened to be circular and sustainable. Tenant engagement differed according to existing approaches and organisational structures; the type of circular project (renovation or new development); and ranged from informational knowledge-transfer type engagement (traditional and consumerist) to more collaborative and co-designed processes (citizenship-based engagement). All four case studies incorporated business-as-usual approaches but extended these to new circular engagement with their tenants.

The case study SHOs aspired to and embraced a sustainable housing agenda and saw CE as an organisational-wide vision or ethos, but there was no expectation or specific plan to encourage tenants to adopt this vision and ethos too, beyond accepting and understanding the different materials and designed systems, how it would impact them and what they would need to do differently in their homes to adapt to those different materials and systems. This was understandable as tenants of social housing are often socially and economically disadvantaged, and all the case study SHOs were apprehensive about the burden of over-consultation on quite technical changes to the tenants' properties.

Interestingly, the new circular tenant engagement did not always reflect the same type of approaches adopted before; for example, Case Study 1 had a cooperative organisational model but for their circular project it was not possible to set up a Tenant Design Board because of the constraints of the innovative but limited material choices to replace traditional plastic components. For Case Study 3, a very collaborative design process with tenants informed the project development but delays in the project development meant ultimately those tenants moved into alternative properties rather than wait for years to take up tenancy in the circular homes they helped design.

Ultimately all case studies focused on the economic and environmental output for their circular homes and considered that the social dimension was addressed by the nature of the product, i.e., high-quality, affordable social homes for their tenants. Key tenant engagement implications for the case studies included gaining necessary tenant acceptance for the circular schemes (particularly refurbishments) as well as informing tenants about any technical changes as to how their circular homes would differ from any previous homes.

Table 4.2 compares and contrasts these business-as-usual and new circular engagement processes evident through the case study findings.

Conclusion

SHOs have an entrenched social purpose serving disadvantaged communities and therefore were considered ideal case study candidates to explore the "missing social link" of circularity. However, our study has revealed that because of the clear social impact and purpose of these organisations, there is an assumption of implicit social impact through their circular housing schemes. Indeed, as their tenant engagement with these circular projects illustrates, the aim for the case study SHOs was to provide an equally safe, comfortable, and affordable home without the ecological aspects of their circular homes being a barrier or burden to tenants. Furthermore, the projects lacked formal mechanisms and the resource to capture the social value of the circular housing projects compared to clear environmental and financial monitoring imperatives. This was understandable in the context of the case studies who were fulfilling their funding obligations and project outputs around the material aspects of their circular development and asset management projects.

Regarding tenant engagement, there were organisational scale and structural factors that influenced the type of engagement that was possible and whether the case study SHOs were able to continue with or newly adopt cooperative and co-design approaches, as some did. All the cases studies adopted business-as-usual tenant engagement and participation. New circular tenant engagement was, however, also needed to gain tenant acceptance for the proposed circular schemes, particularly those that involved refurbishment to existing properties with sitting tenants. New circular engagement also meant reduced choice around materials and design because of technical material, construction, and planning constraints.

Table 4.2 Business-as-usual and new circular tenant engagement in circular social housing case studies

Business-as-usual tenant engagement	New circular engagement	Changes in tenant engagement approach
A good quality, affordable home provided or refurbished – the circularity of the home is incidental	Knowledge transfer – informing and educating tenants on any necessary technical changes to how they should occupy and use their home	Consumerist → informational (Cases 1, 2, 3, 4)
Limited impact on lifestyle and values	Some behavioural change required for sustainable living in social circular homes but no expectation for tenants to take on circular ethos and values that case study organisations had adopted	Traditional → "light-touch" inclusivity (Cases 1, 3, 4)
Tenant consultation/ participation in design and choice of finishes and fixtures	Cooperative, co-design, and collaboration models with constraints – reduced involvement and choice because of technical priorities, sustainable material parameters, and planning constraints	Citizenship + consumerist → constrained collaboration (Cases 1, 4)
Case study organisations' tenant boards, committees or associations for regulatory scrutiny and governance	Limited specific governance and decision-making for tenants around the circular homes specifically, over and above existing governance and engagement structures	Citizenship > some participatory decision making (Cases 1, 3)
Tenant acceptance and agreement on any refurbishment programme as per existing engagement structures	Further specific engagement and reassurance needed especially to overcome barriers to reused materials and component parts for circular projects	Consumerist → enhanced customer service/ satisfaction rationale (Cases 1, 2, 3, 4)

Note: Developed by Sacranie and Cetin, drawing on Cairncross et al.'s 1994 typology of tenant engagement.

It is critically important to note that this study coincided with the COVID-19 pandemic, which delayed the circular housing projects and affected their "business-as-usual" tenant engagement. The SHOs were in crisis management mode during this period and sustaining tenancies and tenant well-being were regarded as more of a priority than engaging tenants on circular developments and renovations.

It can be argued that circular social housing is presently a top-down organisational strategy of adopting a sustainable approach to housing construction and renovation, and this ties into more traditional landlord approaches of making choices for, or on behalf of, their tenants or customers in their best interests as not-for-profit organisations. To embed more of a social solidarity model in circular social homes, it would require SHOs to move beyond implicit social impacts by nature of the service provided, to circular models for housing that build in social indicators, outputs, and mechanisms that begin to redress the critical missing social dimension of the circular economy. Moving towards or reinforcing existing collaborative and inclusive tenant engagement would mean including tenants in the decision-making processes around adopting circular and sustainable principles and values to development and refurbishment, which projects to take on and how to deliver them. More inclusive approaches beyond quality assurance, customer satisfaction, and knowledge transfer about new components would afford the opportunity for tenants and communities collectively to adopt beneficial, sustainable lifestyle choices living in circular homes and neighbourhoods (at the same time as their social housing providers have embraced the principles and values of CE as corporate strategy).

Despite the challenges and constraints of working in a pioneering technical context, if these collaborative models of tenant engagement were strengthened rather than limited by circular strategies, it would allow circularity to become truly entrenched in these organisations and their housing stock, economically, environmentally, and socially through their tenants.

It would be remiss not to acknowledge too that the case study SHOs described in this chapter were at the early development stages of the circular housing projects that were the subject of this research study. While organisational viewpoints have been captured through documentary evidence and interviews, the social aspect of circular social homes and how tenants have been engaged cannot be explored comprehensively without gathering evidence from tenants once they move into their new homes, or have their circular refurbishments completed in their existing social rented homes.

This will constitute the next phase for this ongoing research with the aim to gain further understanding and insight into lived experiences, how tenants have been engaged, the social impacts of their circular homes, and if and how their behaviour and attitudes change by living in a more sustainable dwelling. Examples of behavioural change might range from what is required and necessary to adapt to living in a circular, sustainable home to more aspirational culture, ethos, and lifestyle shifts where communities feel empowered, engaged, and confident to make greener environmental choices along with their housing provider that could have positive economic and social impacts on their lives. Finally, for circular economic schemes like the social housing ones in this research study, the neglected social dimension of CE can begin to be addressed if social inclusion and social impact are built into the project from the design stage through to the project delivery, capturing output and impact beyond economic and environmental targets.

Notes

1 Social housing organisations (SHOs) are also called registered social landlords (RSLs), social landlords, housing corporations, housing associations, and social housing providers.
2 This was the initial phase of the circular projects involving the design, planning, and engaging with tenants about the development/refurbishment works to follow. In the

following stages of this research study, we hope to gather evidence from tenants once they have moved into their homes to investigate the social and lived-in impact of these circular housing schemes.

3 www.nweurope.eu/projects/project-search/charm-circular-housing-asset-renovation-management/

References

AEDES. 2016. *Dutch Social Housing in a Nutshell*. AEDES, The Hague, Brussels

Arnstein, Sherry R. 1969. "A ladder of citizen participation." *Journal of the American Institute of Planners* 4: 216–224 https://doi.org/https://doi-org.tudelft.idm.oclc.org/10.1080/01944366908977225.

Berger-Schmitt, Regina 2002. "Considering social cohesion in quality of life assessments: Concept and measurement." *Social Indicators Research* 58, no. ssessing National Quality of Life and Living Conditions (1): 3–428.

Bocken, Nancy M. P., Ingrid de Pauw, Conny Bakker, and Bram van der Grinten. 2016. "Product design and business model strategies for a circular economy." *Journal of Industrial and Production Engineering* 33 (5): 308–320. https://doi.org/10.1080/21681015.2016.1172124.

Cairncross, L., D. Clapham, and R. Goodlad. 1994. Tenant Participation and Tenant Power in British Council Housing. *Public Administration* 72: 177–200. https://doi.org/10.1111/j.1467-9299.1994.tb01007.x

Çetin, S., V. H. Gruis, and A. Straub. 2021. "Towards circular social housing: An exploration of practices, barriers, and enablers." *Sustainability* 13 (4). https://doi.org/10.3390/su13042100.

Eikelenboom, Manon, Thomas B. Long, and Gjalt de Jong. 2021. "Circular strategies for social housing associations: Lessons from a Dutch case." *Journal of Cleaner Production* 292. https://doi.org/10.1016/j.jclepro.2021.126024.

EMF. 2012. "Towards a circular economy: Business rationale for an accelerated transition." Accessed 27 August 2021. https://ellenmacarthurfoundation.org/towards-a-circular-economy-business-rationale-for-an-accelerated-transition.

European Commission. 2013. "EU buildings factsheets." https://ec.europa.eu/energy/eu-buildings-factsheets_en.

———. 2020. "In focus: Energy efficiency in buildings." Accessed 27 August 2021. https://ec.europa.eu/info/news/focus-energy-efficiency-buildings-2020-feb-17_en.

Eurostat. 2018. "Waste statistics." Accessed 27 August 2021. https://ec.europa.eu/eurostat/statistics-explained/index.php?title=Waste_statistics#Total_waste_generation.

Forrest, Ray, and Ade Kearns. 2016. "Social cohesion, social capital and the neighbourhood." *Urban Studies* 38 (12): 2125–2143. https://doi.org/10.1080/00420980120087081.

Geißdörfer, M., P. Savaget, N. Bocken, and E.-J. Hultink. 2017. "The Circular Economy: A new sustainability paradigm?." *Journal of Cleaner Production* 143: 757–768. https://doi.org/10.1016/j.jclepro.2016.12.048.

Granath Hansson, Anna, and Björn Lundgren. 2018. "Defining social housing: A discussion on the suitable criteria." *Housing, Theory and Society* 36 (2): 149–166. https://doi.org/10.1080/14036096.2018.1459826.

Herczeg, Márton, David McKinnon , Leonidas Milios, Ioannis Bakas, Erik Klaassens, Katarina Svatikova, and Oscar Widerberg 2014. *Resource Efficiency in the Building Sector*. Rotterdam: Ecorys & DG Environment.

Hickman, Paul, and Jenny Preece. 2019. *Understanding Social Housing Landlords' Approaches to Tenant Participation.* Glasgow, Scotland: UK Collaborative Center for Housing Evidence.

Housing Europe. 2021. *The State of Housing in Europe 2021*, Housing Europe, Brussels

Kirchherr, Julian, Denise Reike, and Marko Hekkert. 2017. "Conceptualizing the circular economy: An analysis of 114 definitions." *Resources, Conservation and Recycling* 127: 221–232. https://doi.org/10.1016/j.resconrec.2017.09.005.

Korhonen, Jouni, Antero Honkasalo, and Jyri Seppälä. 2018. "Circular economy: The concept and its limitations." *Ecological Economics* 143: 37–46. https://doi.org/10.1016/j.ecolecon.2017.06.041.

Marchesi, Marianna, and Chris Tweed. 2021. "Social innovation for a circular economy in social housing." *Sustainable Cities and Society* 71. https://doi.org/10.1016/j.scs.2021.102925.

Merli, Roberto, Michele Preziosi, and Alessia Acampora. 2018. "How do scholars approach the circular economy? A systematic literature review." *Journal of Cleaner Production* 178: 703–722. https://doi.org/10.1016/j.jclepro.2017.12.112.

Moreau, Vincent, Marlyne Sahakian, Pascal van Griethuysen, and François Vuille. 2017. "Coming full circle: Why social and institutional dimensions matter for the circular economy." *Journal of Industrial Ecology* 21 (3): 497–506. https://doi.org/10.1111/jiec.12598.

Mullins, David, Peter Shanks, and Halima Sacranie. 2017. *Tenant Involvement in Governance: Models and Practices.* Birmingham: University of Birmingham.

Munaro, Mayara Regina, Sérgio Fernando Tavares, and Luís Bragança. 2020. "Towards circular and more sustainable buildings: A systematic literature review on the circular economy in the built environment." *Journal of Cleaner Production* 260. https://doi.org/10.1016/j.jclepro.2020.121134.

Murray, Alan, Keith Skene, and Kathryn Haynes. 2015. "The circular economy: An interdisciplinary exploration of the concept and application in a global context." *Journal of Business Ethics* 140 (3): 369–380. https://doi.org/10.1007/s10551-015-2693-2.

Padilla-Rivera, Alejandro, Sara Russo-Garrido, and Nicolas Merveille. 2020. "Addressing the social aspects of a circular economy: A systematic literature review." *Sustainability* 12 (19). https://doi.org/10.3390/su12197912.

Pawson, Hal, Janis Bright, Lars Engberg, Gerard van Bortel, Laurie McCormack, and Filip Sosenko. 2012. *Resident Involvement in social housing in the UK and Europe.* Edinburgh: Heriot Watt University.

Sacranie, Halima. 2012. "Hybridity enacted in a large english housing association: A tale of strategy, culture and community investment." *Housing Studies* 27 (4): 533–552. https://doi.org/10.1080/02673037.2012.689691.

Woodard, Ryan, and Anthea Rossouw. 2021. "An evaluation of interventions for improving pro-environmental waste behaviour in social housing." *Sustainability* 13 (13). https://doi.org/10.3390/su13137272.

Yin, Robert K. 2018. *Case Study Research and Applications: Design and Methods.* Los Angeles: SAGE Publishing.

5 How does the circular economy interact with urban food system transformation?

Carlos Miret

Introduction

The global food system is a clear reflection of the problems of the linear economy: it drives land-use change, biodiversity loss, pollution of ecosystems, depletion of natural resources, and climate change, pushing the ecological limits and impacting humanity's life support systems (Springmann et al. 2018; Rockström et al. 2009). Heavily dependent on fossil fuels, the current linear approach to deploying systems of provision (i.e., food, water, waste management, energy) takes for granted the unlimited availability of energy and raw materials and the infinite capacity of the environment to absorb pollution and waste (Jones, Pimbert, and Jiggins 2011). Besides risks to the environment, this approach to food production and consumption compromises human health: whilst over 820 million people are malnourished, the increasing intake of unhealthy foods generates diet-related diseases (Willett et al. 2019).

Within the context of rapid urbanisation, urban areas are increasingly recognised as strategic sites to advance global sustainability by responding to these complex socio-ecological issues (Sonnino, Tegoni, and De Cunto 2019). The interplay between global trends associated with the agri-food system (e.g., vulnerability to food price hikes, increase of food insecurity, climate change impacts on food production, etc.) depicts a "new food equation" in which the *multifunctionality* of food is acknowledged to have a critical role in shaping human health and well-being but also natural resources, animal treatment, and land use (Morgan 2015; Morgan and Sonnino 2010).

In the open-ended and messy process of governing sustainability transitions (Meadowcroft 2009), both the CE and food strategies emerge as promising levers for sustainable development at an urban level (Ilieva 2017; Schroeder, Anggraeni, and Weber 2018). These fields share the need to implement systemic approaches that involve many different actors across scales (Fassio and Minotti 2019; IPES-Food 2017). But little is known about their interaction in the transition to a sustainable urban food system.

The attractiveness of the CE for businesses and governments has motivated the adoption of CE programmes in many cities (Campbell-Johnston et al. 2019; Jones and Comfort 2018), as a new strategy to achieve sustainable goals (Sodiq et al.

DOI: 10.4324/9781003255246-5

2019; De Jong et al. 2015). However, several scholars (e.g., Geissdoerfer et al. 2017; Fratini, Georg, and Jørgensen 2019) highlight that the popularisation of this "go-to concept" (Calisto Friant, Vermeulen, and Salomone 2020: 1) is wrapped in a scrim of ambiguity: employed inconsistently by many stakeholders from different backgrounds (Preston 2012), different discourses arise with potential competing interests. Furthermore, the lack of attention paid to the social dimension of sustainability contests the CE as a promising tool to attain Sustainable Development Goals (SDGs) targets (Schroeder, Anggraeni, and Weber 2018). The impacts of the CE on social equity and future generations are barely mentioned by scholars (Kirchherr, Reike, and Hekkert 2017) and most CE literature only addresses social well-being in association with the creation of jobs (Geissdoerfer et al. 2017). This raises important questions for the transformation of urban food systems. For example, is the mainstream CE frame fit for addressing persistent socio-economic issues that drive food insecurity and malnutrition?

In this chapter, I take the case study of the English city of Brighton and Hove (B&H) to explore the role of an urban food strategy in the development of a circular food system. For this purpose, I identify local initiatives which are key in satisfying CE for food ambitions. By doing so, this case study sheds light on the important role that food initiatives in the social and solidarity economy (SSE) play in building community resilience and in enacting CE principles.

Through a transdisciplinary approach that integrates urban food governance, urban food policy, sustainable cities, food poverty, and food waste, I review literature on the CE to interrogate the application of this concept in urban food systems transformation. In the analysis of perceptions around the CE, I identify contrasting narratives and solutions. A discussion on surplus food redistribution as a strategy to reduce food waste and food poverty highlights a caveat: whilst the CE holds potential as a new frame for food system change and cross-sectoral governance, a narrow focus on food waste reduction targets risks neglecting the need for structural changes.

A recently developed conceptual framework that places human development at the core of CE transformation highlights the importance of inclusive business models and the active role of CE community initiatives in building resilience, when enabled by supportive macro-economic policies (Schröder, Lemille, and Desmond 2020). Throughout the chapter, I use this framework to identify social and cultural opportunities for food system transformation which go beyond the cycling of resources. I also use Schulz, Hjaltadóttir, and Hild's (2019) broad interpretation of the CE to deliberately include the social dimension of circular practices and to emphasise the role of actors that sit outside of the formal organisations of the market economy. I argue that these play a crucial role in local food transformation (e.g., grassroots innovations) (Gernert, El Bilali, and Strassner 2018) but are often absent in the mainstream CE policy frame (e.g., EU policies, major circular city strategies, corporate plans, etc.).

This chapter is structured as follows: after this introduction, I describe the context of the case, followed by a brief review of food policy and CE strategies at the city level. The principles of circular food systems are outlined in the following section, and supported by findings from a mapping of B&H initiatives against CE for food ambitions. The chapter continues with a discussion on the relationship between contrasting CE narratives and the paradigm of unlimited economic growth. In the next section, I focus on the localisation narrative as contestation to the linear economy. This is followed by a discussion on surplus food redistribution as a win-win for food waste and food poverty within the CE frame. Before the concluding remarks, I discuss the role of socially driven food organisations in the transition to sustainable food systems and their interaction with the CE.

Case study context

Brighton and Hove represents a municipality in transition to the CE, with an *Economic Strategy* (BHCC 2018) that explicitly frames the CE as a tool for sustainable economic growth. Under the umbrella of the Brighton and Hove's Food Partnership (BHFP), this city pioneered a systemic approach to food system transformation through a collaborative effort between public, private and voluntary organisations (Sonnino 2019). Located on the South-East English coast and with 300,000 residents (BH Connected 2017), B&H fosters great civic engagement and suffers from several pressures: a high population density and rapid urbanisation growth; one of the lowest levels of housing affordability in the UK; severe inequality; persistent issues of homelessness, poverty, and mental health (Ehnert et al. 2018; BHCC 2018). The rise of demand for foodbanks reflects issues of food insecurity and malnutrition which disproportionately affect the city's poorest (BHFP 2018).

To produce a rich knowledge of the case, I carried out a thematic analysis (Braun and Clarke 2006) of primary data from 22 semi-structured interviews with local CE and city-region food system stakeholders, and secondary data from key policy documents and food-related local case studies (e.g., public food procurement, food poverty). Participants were shortlisted based on these conditions: a) representatives from the BHFP; b) key developers of the city's CE Roadmap; and c) partner organisations in the *Food Action Plan* (B&H's food strategy), the actions of which aligned with one or more CE for food ambitions.

Governing sustainable cities through the lens of food and the CE

The cross-sectoral nature of food directly or indirectly links all the SDGs (Rockström and Sukhdev 2016). Comparative analyses between food strategies and the 2030 Agenda of cities in the Global North (Ilieva 2017; Olsson 2018) demonstrate the potential of using food as a vehicle for delivering SDGs at a local level. A conclusion from these studies is that cities are key in planning the transition towards sustainable and just food systems, as much as food system sustainability is vital to attain the sustainable city vision (Ilieva 2017).

Traditionally, the interconnections between food and other resources (i.e., water and energy) and between food and other sectors (e.g., housing, transport, environment, etc.) have been poorly addressed by governments; these issues have been siloed in different policy areas (Barling, Lang, and Caraher 2002; Sonnino, Tegoni, and De Cunto 2019). Although the paradigm of a "corporatist producer-driven" alliance between national governments and international bodies has dominated the food policy realm in most countries, the last decades have seen the proliferation of new food governance systems at the local level (Morgan 2015). These are new "spaces of deliberation" in the form of food policy councils, food boards, and food partnerships (Moragues et al. 2013) which "constitute a meeting place for civil society, private actors, and the local state to transition towards more just and sustainable urban food systems" (Moragues-faus and Morgan 2015: 1559). More recently, the need to integrate different policy areas to address food system issues (IPES Food 2019) has motivated a new transdisciplinary scholarship ("CE for food policy") that advocates using the CE as a tool to devise holistic strategies (Fassio and Minotti 2019).

Cities have increasingly become places for food policy innovation through the adoption of systemic approaches to food sustainability (Sonnino, Tegoni, and De Cunto 2019). Many UK cities, for example, have developed urban food strategies to address the socio-ecological problems that stem from the industrial food system (Morgan 2015). These strategies integrate different areas of action in which food has an impact (e.g., health, the environment, community development, etc.), including deep-rooted issues such as food poverty (Moragues et al. 2013). Driven by a collaboration between the BHFP, the BHCC, and local organisations, B&H's *Food Poverty Action Plan* constitutes an example of an urban food policy which strives for long-term social, educational and public health benefits, by focusing on actions that tackle the underlying causes of the problem (e.g., access to employment, cooking skills, affordable housing, etc.) (BHFP 2015a).

Building on the CE momentum, many European cities have developed CE roadmaps with reports that demonstrate contextual efforts to develop CE strategies for specific urban systems; e.g., Glasgow (Circular Glasgow 2016), Paris (Mairie de Paris 2017). Despite increasing attention paid to urban CE strategies in recent literature (e.g., Prendeville, Cherim, and Bocken 2018; Gravagnuolo, Angrisano, and Girard 2019), the lack of an evidence-based theoretical framework means that CE implementation at the macro-level (i.e., cities, regions) (Velenturf and Purnell 2021) is mainly led by practitioners (Sánchez Levoso et al. 2020), which commonly regard the CE only from a waste or environmental management perspective. For example, case studies from Dutch cities show that, within the broader EU CE policy framework, which aims to maximise resource recovery, municipalities focus on end-of-cycle actions, partly due to the lack of instrumental capacity to influence value chains (Campbell-Johnston et al. 2019).

The issue of scale has been highlighted as one of the common challenges to implementing systemic change in cities (Sonnino, Tegoni, and De Cunto 2019), since structural problems originate at broader scales (IPES-Food 2017). For example, the disconnection between EU, national, and local authorities poses barriers to harnessing the potential of alternative food system initiatives (e.g., community-supported agriculture, urban agriculture, short supply chains, etc.) as

enablers of a transition towards sustainable food systems; so far these urban innovations remain niches in food supply markets (IPES Food 2019).

Circular food systems

Although there is no clear conception of a circular food system, the Ellen
MacArthur Foundation (EMF), a key organisation in building the CE momentum (Lazarevic and Valve 2017), and whose reports are widely used to guide
policymakers (Williams 2019), advocates for tapping the following levers (Ellen
McArthur Foundation 2015): increasing resource-efficient agricultural practices
through technological solutions (e.g., precision agriculture); embracing regenerative farming practices (e.g., no-till, agroforestry, holistic-planned grazing, etc.);
closing nutrient and material loops (e.g., diverting urine for fertiliser, composting bio-waste, or producing bio-energy through anaerobic digestion); restoring
and preserving natural capital; promoting urban and peri-urban farming to reduce
"food miles" and strengthen links between rural and urban areas; and enhancing
digital supply chains to reduce food waste.

Similarly, Jurgilevich et al. (2016) identified three stages in the food system to
implement circular strategies:

- Production: focusing on closing nutrient loops and supporting local farming.
- Consumption: shifting towards more plant-based diets and educating consumers on food waste minimisation.
- Food waste and surplus management: taking a preventative approach to food
  waste; redistributing surplus for human consumption; adding value to waste
  by cascading materials, generating compost or biogas.

Improving the connectivity of resource flows is considered one of the guiding
principles to develop the resilience of urban food systems (Wiskerke 2015) and
their rural surroundings, which provide them with ecosystem services (Herbert
Girardet 2017). Within the scale of city-region, opportunities for CE actions
increase through rural–urban linkages that facilitate organic wastes recycling and
the management of resource flows (Dubbeling et al. 2016). Drawing from circular
metabolism or "cradle-to-cradle" approaches, there is a range of context-specific
strategies to revalorising waste: from centralised, high-tech systems (e.g., agro-
industrial parks based on industrial ecology principles), to decentralised, low-tech
systems (e.g., turning household waste into compost in agroecological production) (Wiskerke 2015).

The *Cities and the Circular Economy for Food* report (Ellen MacArthur
Foundation 2019) comprises one of the only attempts to operationalise the
concept of circular economy for food in cities, based on three comprehensive
ambitions (Table 5.1). The following subsections describe some of the findings
from a non-exhaustive mapping of local food initiatives (embedded in B&H's
food strategy) against the three CE for food ambitions, which is summarised
in Table 5.2.

Table 5.1 Three CE for food ambitions and main proposed actions

Ambitions	Proposed actions
1) Source food regeneratively and locally grown, where appropriate	• Using demand power to influence the adoption of regenerative agricultural practices • Enhancing urban farming and embracing the connection with peri-urban farms • Taking advantage of technologies and innovation to improve food production and local supply chains (e.g., organic fertilisers, traceability technologies, AI for farm management, ecological footprint trackers, etc.)
2) Make the most of food	• Redistributing surplus food for human consumption • Generating value from organic waste collection (e.g., bioenergy) with uses in food production (e.g., compost, fertilisers) • Repurposing food by-products to create diverse new products
3) Design and market healthier food products	• Design food products to be safely repurposed for new uses • Use marketing to promote plant-based diets and the use of regenerative ingredients

Note: adapted from Ellen MacArthur Foundation (2019).

Source regeneratively and locally where appropriate

There are 75 community gardens across the city which enhance qualitative values such as mental well-being and social cohesion, but they are a niche practice in urban food production (BHFP 2018). Several respondents pointed out land as a limiting factor both for CE infrastructure (e.g., repair hubs) and for scaling up urban food production. The unavailability and unaffordability of land are some of the reasons why the traditional role of market gardens and smallholdings in supplying the local food sector has been eroded (Food Matters 2011). Despite the Brighton and Hove City Council (BHCC) owning 4,400 hectares of farmland, these are not used to supply food for the city (BHFP 2018). In this respect, the strategic work of the BHFP to support urban food production includes influencing food policy. For example, in collaboration with the BHCC, the BHFP produced a *Planning Advice Note on Food Growing and Development* (Food Matters and BHCC 2011) to guide and promote the inclusion of food growing sites in new developments.

Historically, the city council the local food sector as a significant contributor to the local economy (Food Matters 2011). However, the prominent tourist economy is a focal entry point for implementing circular actions (e.g., the influential power of the hospitality sector to favour sustainably-grown ingredients and plant-based menus; opportunities for repurposing big volumes of food waste, etc.). An example of current potential was found in plans for the development of a local anaerobic digestion plant, which aims to collect food waste from restaurants to be transformed into biogas for a community energy model.

Table 5.2 Mapping of local food initiatives against three ambitions for an urban CE for food

Ambitions	Initiatives
Source food regeneratively and locally grown where appropriate	**Food production** • Urban food production: allotments, community gardens, community orchards • Peri-urban agriculture: community-supported agriculture • Regional food production: local fisheries, Sussex farmers, small-scale organic vegetable box-schemes **Local ecosystem** • Bodies for sustainable management and restoration of land and marine environment **Policy** • Planning: e.g., *Planning Advice Note* on food growing and development • Procurement: e.g., *Minimum Buying Standards* **Food consumption** • Farmer's markets, farm to fork retail • Local food distributors **Regenerative agriculture** • Facilitating land access for new entrants to small-scale agroecological farming
Make the most of food	**Surplus food redistribution** • Surplus Food Network **Surplus fruit products** • Community circular food businesses (e.g., juice producers from surplus apples from community orchards) **Surplus food community meals** • Shared meals (e.g., lunch clubs) • Community-owned pub **Repurpose food waste** • Food waste to community biogas (i.e., local anaerobic digestion plant) • Public food waste collection (household and businesses) • Community composting schemes • On-site composting (e.g., universities)
Design and market healthier food products	Promotion of healthy and plant-based diets: • Hospitality • Public procurement • National/city campaigns

Source: author.
Notes: the table shows current initiatives in the B&H city-region food system which enact CE principles or which are considered key to achieving the ambitions reported on the cities and circular economy for food report (Ellen MacArthur Foundation 2019). Initiatives were identified through interviews and desk-based research.

The B&H case illustrates how public food procurement emerges as an important tool to overcome the challenge of scale. Through the coordination of the Good Food Procurement group, the BHFP acts as a hub that provides networking and workshops, showcases best practices and liaises with local suppliers (BHFP 2015b). This group facilitated the adoption of "Minimum Buying Standards" (based on the UK's Soil Association's *Bronze Food for Life*) for all of the BHCC catering contracts (BHFP 2016). The collaboration between the public sector, citizens, and commercial and voluntary organisations fostered the School Meals Service: a city-wide single contract which contributed to reducing inequalities in the access to healthy diets between rich and poor children; sourced ingredients from local farms committed to environmental sustainability standards; and strengthened the local economy through fair employment (BHFP 2017). Previous studies about school meal reforms show that, even when embedded in neoliberal power dynamics, innovative public food procurement from the sub-national state is a form of local empowerment that produces sustainable development outcomes (Sonnino 2010).

Make the most of food

Maximising surplus food redistribution through the coordination of the Surplus Food Network[1] is an important focus of the BHFP. In 2018, over 1,090 tonnes of surplus food were intercepted by this organisation; an 87% increase since the network was formed in 2016 (Starr-Keddle 2019).

Other actions to make the most of food include community composting. Coordinated by the BHFP, the city now has 37 schemes which enable over 1,000 households to compost their food waste, which is then used as fertiliser in community gardens. However, this service is constrained by stable funding.

The BHFP has worked collaboratively with the BHCC to develop a "food-use" citywide strategy based on the food waste hierarchy. For example, the BHFP carried out a food waste audit in primary schools with the aim to identify the areas that lead to food waste:

> A lot of the food being wasted was vegetables, which is interesting, because that kind of focus came from projects around getting people to eat more vegetables. And the School Meals Service has a legal obligation to meet certain standards, which means they have to put a certain amount of vegetables on the plate, quite rightly, to encourage children to eat vegetables, but if a lot of those are ending in the bin, that's not benefitting anyone. And that's a bit of policy not helping with another bit of policy.
>
> (BHFP Representative)

This example sheds light on the need for preventative approaches to food waste. It also suggests that cross-departmental and policy integration are required to avoid unintended trade-offs between different areas (i.e., promoting healthy diets and food waste reduction), which supports findings from case studies on urban food policies across the world (IPES-Food 2017).

Design and market healthier food products

Whilst local governments and organisations have little influence on the city's food supply, they can promote the consumption of plant-based diets and healthy products. Partnered with national organisations like Sustain and The Food Foundation (the latter is the national lead for campaigns such as "Peas Please"[2]) the BHFP and the BHCC made a joint pledge to help residents eat more vegetables under the citywide "Veg City" campaign. This is one example that contributes to promoting the consumption of healthy and plant-based diets; through education, communication, and training, an array of partnered organisations (e.g., schools, the BHCC, Sussex Community NHS Foundation Trust, etc.) work on improving the nutrition of residents, including facilitating access and affordability of healthy diets for people in situations of food poverty (BHFP 2018).

Contrasting narratives for a new economy

Chaturvedi, Gaurav, and Gupta (2017) argue that transitioning to a circular economic system is "a political process" driven by narratives that reflect the ideological positioning of different actors across scales. Different narratives are underpinned by worldviews about the economy–society–nature relationship in which unlimited economic growth is either supported or contested, and which might lead to more incremental or radical actions (Temesgen, Storsletten, and Jakobsen 2019). Visions that maximise resource recovery through technological infrastructure, for example, contrast with sufficiency approaches that prioritise waste prevention, reusing, and repairing (Velenturf and Purnell 2021).

The B&H case serves as an example of how CE consultants, policymakers, and a diverse group of food system stakeholders can produce contrasting visions of the CE. Through 22 interviews, I identified a dominant narrative which portrays the CE as an economy in which waste is revalued. This narrative, which I call "zero-waste cycles", embodies the origins of the concept in industrial ecology and its evolution into zero-waste policy discourses (Preston 2012).

A few participants in this study associated the CE with the positive impacts of local circulation of resources and expenditure within the local economy. For example:

> The food that is grown locally and sold locally is overall more beneficial to the local economy, because you are funding the farms, you are funding the employment...that money is then being spent within the communities... the circularity of expenditure makes it beneficial for the whole community.
>
> (Distributor of local produce to organisations fighting food poverty)

This second narrative, which I call "local CE", sits within a broader narrative that features the local economy as the key to break away from the lock-ins of a globalised economy; it reproduces the *eco-localism* paradigm (Curtis 2003) and echoes the Transition Town movement, which promotes building community resilience by closing economic loops within the local economy (Hopkins 2008). According to Longhurst et al. (2016) this "counter-narrative" is underpinned by *degrowth* approaches which deliberately critique global capitalism and the pursuit of economic growth as the basis for economic development.

Compared to approaches like *degrowth* or a steady-state economy, the positive framing of the CE discourse has built momentum for societal change amongst policymakers and the corporate world, partly because it emphasises economic benefits (Geissdoerfer et al. 2017). The CE as a zero-waste paradigm has been institutionalised through a mainstream discourse that fails to challenge the basis of unlimited economic growth, inherently supporting the status quo of the linear economy (Hobson and Lynch 2016; Temesgen, Storsletten, and Jakobsen 2019). Some scholars argue that the current CE framing reproduces the "ecological modernisation" discourse: focusing too much on the role of businesses and smart technologies; attracting incumbent actors (e.g., global corporations) to lead the transition; failing to address the roots of environmental damage; and perpetuating capitalist norms (Hobson and Lynch 2016).

Whilst different discourses around the CE might be classified in terms of "weak" or "strong" circularity (Johansson and Henriksson 2020), some scholars argue for integrating the overlapping elements of mainstream CE and *degrowth* approaches (e.g., product durability, shrinking of extractive industry, etc.) in order to build a strong alternative to the incumbent linear model. For example, short supply chains and low-input farming practices are solutions both shared by *degrowth* and mainstream CE discourses. However, whilst the first emphasises local food self-sufficiency the latter promotes technological solutions to decouple resource consumption from economic growth. Circular strategies of major cities tend to focus on the role of smart technologies and digital innovation in enabling the efficiency of supply chains, and the development of infrastructure that allows high-value recovery of organic residual streams (e.g., Circle Economy, Fabric TNO, and Gemeente Amsterdam 2016; Circular Glasgow 2016). These solutions, which align with the technocratic "smart city" concept (Maye 2019), contrast with visions of circular food systems in which the CE is associated with agroecological approaches that promote fair and sustainable food chains (FAO 2018). Some of these include small-scale, low-tech, integrated systems of energy, water and waste which are advocated as localisation models that build resilient livelihoods against global externalities (Jones, Pimbert, and Jiggins 2011; Albala and Pimbert 2015).

Localisation: an antidote to the linear economy?

The feasibility and the impacts on society and the environment of pursuing alternative pathways to globalised food supply chains remain highly unknown (Gomiero 2018). But, amidst frequent and deep global crises, the need to develop local resilience is pushing for an increased localisation of supply and production systems, which could benefit from CE solutions (Sarkis et al. 2020). At the same time, some scholars argue that the transition to a CE in a city region requires unlocking the potential of existing local initiatives and strengthening the local economy (Nogueira, Ashton, and Teixeira 2019).

Increasing local food production and consumption has been advocated by many as a strategy to confront the unsustainability of the global food system (Sonnino 2010). Short supply chains could contribute to the CE through potential minimisation of food waste (e.g., less packaging, promotion of environmental behaviour, etc.) and reduced carbon emissions from shorter transport (Kiss, Ruszkai, and Takács-György 2019). A plethora of scholars, however, warn about falling into the "local trap"; i.e., making assumptions that localisation equals environmental gains and failing to recognise the potential of local markets in exacerbating social inequalities associated with the access to food (e.g., Hinrichs 2003; Born and Purcell 2004). This is acknowledged by the BHFP, which shifted their initial focus on local food to a broader sustainability strategy (Sonnino 2016) with an emphasis on social justice: "fair food for all" (BHFP 2018: 6). The approach of this partnership signals a new form of localism that defines 'local food' beyond territoriality, founded on values of fairness and solidarity and engaging in multiscalar relationships to support the concept of "good food" (e.g., ethically sourced food, sustainable farming, nutritious diets, etc.) (Sonnino 2019).

Whilst the CE offers a new frame for strategic work on food system change (Fassio and Tecco 2019), findings from the B&H case indicate that most perceptions around the CE for food are mostly associated with food waste minimisation. Although this is not surprising, considering the zero-waste narrative of the CE, this one-dimensional view strips the problem of food waste from its socio-economic context and has potential implications for societal transformation. I discuss this in the next section.

Surplus food: a solution to food waste and food poverty?

In the last decade, fighting the environmental impact of food waste has become the focus for many organisations previously engaged with other food activities (Mourad 2016). This can also be observed in the BHFP, due to several factors: having little power to influence the city's food supply; an urgent need to alleviate food poverty through emergency networks; a context of climate emergency and food waste reduction targets (e.g., the national Courtauld 2025 Commitment[3] and the 12.3[4] target of the SDGs).

In a context of rising demand for food banks and the urgency to address food waste, surplus food redistribution is nationally advocated as a win-win to reduce food waste and solve food insecurity (Caraher and Furey 2017). From a CE perspective, the redistribution of surplus food for human consumption comprises a high-value option, classed as waste prevention within the food waste hierarchy (WRAP 2020; Papargyropoulou et al. 2014). However, advocating the use of surplus as the solution to food insecurity poses the risk to "depoliticise" hunger and allow the government to elude responsibility for structural factors that lead to food poverty (e.g., delays in benefit payments, disparities between incomes and food costs, etc.) (Caraher and Furey 2017; Caplan 2017). An example from Finland warns us that, under EU's CE policy discourse, framing the benefits of surplus food redistribution for environmental sustainability risks reinforcing the institutionalisation of charitable food aid (Tikka 2019).

Organisations that intercept surplus food for poverty alleviation have the potential to create social value for communities (Mirosa et al. 2016) and have been considered a form of social innovation with the potential to transform the food system (Hebinck et al. 2018). They also contribute to achieving the 12.3 target of the SGDs (Schroeder, Anggraeni, and Weber 2018). Whilst the 12.3 is an "efficiency" target aimed at reducing the environmental impacts of food waste, downstream approaches that disregard the transformation of current structures of production and consumption cannot fully realise circular flows of materials (Bengtsson et al. 2018). According to Messner, Richards, and Johnson (2020), the CE contributes to a socio-technical lock-in that promotes the development of technological infrastructure, favouring the management (rather than reduction) of food waste. These arguments are supported by findings from a report (IPES Food 2019) which reveals that current European CE policies on food waste have been geared towards solutions for major retailers, focusing on recycling rather than waste prevention.

Re-thinking the CE for food to include the social dimension

Addressing food system dilemmas (e.g., ensuring affordable food for low-income citizens and fair prices for producers) might require national policy intervention (Anderson and Leach 2019). However, the challenge of breaking away from the legacy of linear mindsets and structures becomes more feasible at the community level (Paiho et al. 2020). Besides the debate about potential environmental gains, there is an argument for localising resource consumption to influence changes in social practices and systems of provision, and therefore the adoption of circular actions (Williams 2019). Indeed, a broad literature on social-ecological systems highlights how a strong cultural identification with place through community-based food practices can bring about virtuous feedbacks of ecosystem management and community development (e.g., Delind 2006; Ostrom 2007; Barthel, Folke, and Colding 2010; Westley et al. 2011; Ravenscroft, Moore, and Welch 2012).

Bottom-up social innovation tends to be overlooked and unsupported in mainstream sustainability policymaking (Bergman et al. 2010). This can be observed in mainstream CE discourses which emphasise technological innovation for business and industry as the enabler of sustainable economic growth (Calisto Friant, Vermeulen, and Salomone 2020; Ziegler 2019). But, technological innovation alone is insufficient to address the inherent complexity of transitioning to a CE, which requires changes in institutional and social practices (de Jesus and Mendonça 2018). Whilst some scholars signal the relevance of the third sector in enacting circular principles in cities (Gravagnuolo, Angrisano, and Girard 2019; Kebłowski, Lambert, and Bassens 2020), respondents of this study see the CE as a platform to build on the key role of B&H's third sector (with 2,300 organisations) in creating local economic and social value (BHCC 2018).

The central role of ecological processes and social interactions in transitioning to sustainable food systems demands non-material innovations that are founded on local knowledge (HLPE 2019) and that see food as the vehicle to safeguard the intrinsic values of sustainability (i.e., "human health, ecological integrity, and social justice") (Moragues-faus and Morgan 2015). In this light, food community projects become an integral part of an urban circular food economy. They constitute important forms of civic and social innovations (Maye 2016) that use food as the vehicle to build community capacity (Kirwan et al. 2014). In the B&H case study, members from different community-driven food organisations (i.e., CSA, community-owned pub, a local gleaning group, surplus food redistributor), pointed at outcomes of "empowerment" and "social agency" from their initiatives; some explicitly associated these to the CE. For example:

> The inclusivity of a community, to me, is a form of CE, by providing what people ask for and what they are prepared to provide for themselves, by volunteering, by helping us create these activities and sometimes entirely taking them on themselves, that is a proper form of sort of neighbourhood, asset based community development, which is hopefully long-lasting, sustainable and truly responsive to the community.
>
> (Community-owned pub manager)

Moreau et al. (2017) suggest that initiatives operating in the Social and Solidarity Economy (SSE) could contribute to CE strategies by integrating principles of solidarity and social equity. These include non-profit and informal activities that contribute to sustainable consumption and zero-waste goals (e.g., edible communities, surplus food sharing, etc.). Often considered in relation to a sharing or collaborative economy (Gorenflo 2017; Cohen and Muñoz 2016), these activities constitute valuable social and cultural interactions outside of market-based networks (Jehlička and Daněk 2017). In the B&H context, a relevant example comprises the projects which offer vulnerable people access to affordable meals in a social setting. In an attempt to move away from a national food poverty agenda focused on foodbanks (Caraher and Dowler 2014), the BHFP found the relevance of "shared meals" (e.g., lunch clubs) in improving the access to nutritional foods, improving mental well-being and reducing social isolation (BHFP 2015a, 2015b). These and other activities of food sharing constitute spaces for social connection that improve the ability of local communities to self-organise and reduce vulnerabilities to food insecurity (Blake 2019).

Concluding remarks

The lack of attention paid to the social dimension of sustainability in mainstream CE discourses poses an important barrier to delivering transformative societal change. This, I argue, is particularly concerning in the context of urban systems where food-related social innovations are key components (Maye 2016) that help address persistent socio-economic issues (e.g., mental health, inequalities in the access to healthy diets, etc.). Socially driven, place-specific food initiatives have the potential to build community resilience and transform social practices. Here, I highlight the need to consider the third sector as an important cog in the city-wide implementation of circular actions. This study highlights land and scale as determining factors for the feasibility of deploying city-wide circular actions and calls for new CE frameworks to consider urban areas as complex socio-ecological systems (Williams 2019).

Although the CE momentum has the potential to attract different actors and develop cross-sectoral collaborations for food system change, under the current policy context the CE appears to be no more than a new frame for optimising resource cycling. As discussed in this chapter, the mainstream CE frame risks institutionalising emergency food practices by emphasising the environmental benefits of surplus food redistribution. In the meantime, cross-scalar socio-political issues that generate waste and drive food insecurity and malnutrition remain unquestioned.

The production and distribution of wastes occur in a global food system which is operated by neoliberal institutions and regulated by social norms across national and international scales (Gille 2012) (e.g., agricultural policies that support monoculture; international trade agreements that facilitate corporate control of value chains; a culture of consumption based on marketing, etc. [Audet and Brisebois 2019]); these are grounded on the prevailing paradigm of continued economic growth and development. I argue that this policy macro-environment perpetuates the linear model and threatens the capacity of communities that engage in diverse SSE food activities to build resilience against external shocks (UNRISD 2016; Blake 2019).

Although SSE initiatives embed the necessary values for a socially inclusive CE (Brown et al. 2020), the "success" of these depends on place-specific factors (Amin, Cameron, and Hudson 2002) which might not be easy to plan for when devising policy goals. In this regard, the CE momentum in policymaking offers new perspectives for systemic transformation (Ávila and Campos 2018). As the B&H case suggests, in cities with a rich civic engagement there is potential for the CE to serve as a new frame to knit together existing sustainability and socially driven initiatives. But this requires an interpretation of the CE that recognises the active role of citizens and communities; one that places importance on social equity, inclusiveness, and diversity. Furthermore, to be a valid tool for food system transformation, I argue that the CE needs to integrate different narratives which prioritise social practices that foster community resilience, rather than prolong resource productivity. But, embedding negative social externalities in a CE approach requires developing qualitative indicators that capture socio-cultural dimensions of the circular urban economy that go beyond energy and material flows (e.g., cultural cooperation) (Gravagnuolo, Angrisano, and Girard 2019).

Supporting an emergent literature on urban food policy (e.g., Moragues-faus and Morgan 2015; Sonnino 2019), this study highlighted the BHFP as a governance arrangement with key attributes to enable a CE transition (Paiho et al. 2020); e.g., capacity to facilitate cross-sectoral collaboration embedded in a long-term strategy; coordination of networks; sharing knowledge and facilitating community development. Other city governments could learn from this example, but more targeted research could contribute to understanding how these attributes can be harnessed to explicitly integrate CE principles and whether they can influence the adoption of business and technical enablers (Paiho et al. 2020).

To sum up, the CE interacts with urban food transformation by providing the guiding principles to attain environmental sustainability from waste minimisation. The momentum around the CE has the potential to give visibility to socially driven food organisations engaged in zero-waste activities, by framing them as key sustainability actors. In turn, food organisations within the SSE add values of equity, inclusion, and solidarity to a CE frame which narrowly focuses on environmental and economic targets.

Notes

1 The Surplus Food Network is a partnership of organisations (including FareShare Sussex, Gleaning Sussex, and The Real Junk Food Project) that work collaboratively to effectively redistribute surplus food across the city to vulnerable residents.
2 The Peas Please initiative aims "to secure commitments from industry and government to improve the availability, acceptability (including convenience), affordability, and quality of the vegetable offer in shops, schools, fast food restaurants and beyond, and in turn stimulate increased vegetable consumption among the UK public, particularly children and those on a low income" (Food Foundation 2020).
3 The Courtauld 2025 Commitment, promoted by the Waste and Resources Action Programme
     (WRAP) is a "voluntary agreement brings together organisations across the food system to make food & drink production and consumption more sustainable" (WRAP 2020).
4 "Halve per-capita global food waste at the retail and consumer levels and to reduce food losses along production and supply chains, including post-harvest losses" (Bengtsson et al. 2018).

References

Albala, Ken, and Michel Pimbert. 2015. "Circular Food Systems." *The SAGE Encyclopedia of Food Issues*, no. February. https://doi.org/10.4135/9781483346304.n81.

Anderson, Molly and Melissa Leach. 2019. "Transforming Food Systems: The Potential of Engaged Political Economy." *The Political Economy of Food* IDS Bulletin 50 (2) July 2019. https://doi.org/10.19088/1968-2019.114.

Amin, Ash, Angus Cameron, and Ray Hudson. 2002. *Placing the Social Economy*. http://books.google.com/books?hl=en&lr=&id=UaLudY3VXn4C&pgis=1.

Audet, René, and Éliane Brisebois. 2019. "The Social Production of Food Waste at the Retail-Consumption Interface." *Sustainability* 11 (14): 1–18. https://doi.org/10.3390/su11143834.

Ávila, Rafael Chaves, and José Luis Monzón Campos. 2018. "The Social Economy Facing Emerging Economic Concepts: Social Innovation, Social Responsibility, Collaborative

Economy, Social Enterprises and Solidary Economy." *CIRIEC-Espana Revista de Economia Publica, Social y Cooperativa* 93: 5–50. https://doi.org/10.7203/CIRIEC-E .93.12901.

Barling, David, Tim Lang, and Martin Caraher. 2002. "Joined-up Food Policy? The Trials of Governance, Public Policy and the Food System." *Social Policy and Administration* 36 (6): 556–74. https://doi.org/10.1111/1467-9515.t01-1-00304.

Barthel, Stephan, Carl Folke, and Johan Colding. 2010. "Social: Ecological Memory in Urban Gardens: Retaining the Capacity for Management of Ecosystem Services." *Global Environmental Change* 20: 255–65. https://doi.org/10.1016/j.gloenvcha.2010.01.001.

Bengtsson, Magnus, Eva Alfredsson, Maurie Cohen, Sylvia Lorek, and Patrick Schroeder. 2018. "Transforming Systems of Consumption and Production for Achieving the Sustainable Development Goals: Moving beyond Efficiency." *Sustainability Science* 13 (6): 1533–47. https://doi.org/10.1007/s11625-018-0582-1.

Bergman, Noam, Nills Markusson, Peter Connor, Lucie Middlemiss, and Miriam Ricci. 2010. "Bottom-up, Social Innovation for Addressing Climate Change." Conference on Energy Transitions in an Interdependent World: What and Where Are the Future Social Science Research Agendas, 1–27. http://www.researchgate.net/profile/Noam_Bergman /publication/228784768_Bottom-up_social_innovation_for_addressing_climate_ change/links/02e7e53a6fd553c253000000.pdf.

BH Connected. 2017. "Health and Wellbeing in Brighton & Hove. Joint Strategic Needs Assessment (JSNA) Executive Summary July 2019." *Brighton & Hove JSNA* July: 1–14.

BHCC. 2018. "Productive, Inclusive, Transformative: An Economic Strategy for Brighton and Hove." 1–80.

BHFP. 2015a. *Brighton & Hove Food Poverty Action Plan 2015–2018.*

———. 2015b. *Good Food Is Good for Business. Case Study: Brighton & Hove's Good Food Procurement Group.*

———. 2016. *Minimum Buying Standards. Brighton & Hove Food Partnership Case Study.*

———. 2017. *Case Study: More than a Meal. How the Brighton & Hove Primary School Meals Service Is Adding Value to Our City.*

———. 2018. *Brighton and Hove Food Strategy Action Plan.*

Blake, Megan K. 2019. "More than Just Food: Food Insecurity and Resilient Place Making through Community Self-Organising." *Sustainability* 11 (10): 2942. https://doi.org/10 .3390/su11102942.

Born, Branden, and Mark Purcell. 2004. "Scale and Food Systems in Planning Research." *Journal of Planning Education and Research* 26: 195–207. https://doi.org/10.1177 /0739456X06291389.

Braun, Virginia, and Victoria Clarke. 2006. "Using Thematic Analysis in Psychology" *Qualitative Research in Psychology* 3 (November): 1–21.

Brown, Esther Goodwin, Luis Sosa, Antonius Schröder, Kris Bachus, and Ödül Bozkurt. 2020. "The Social Economy: A Means for Inclusive & Decent Work in the Circular Economy? Support for This Position Paper From the Circular Jobs Initiative." https:// www.circle-economy.com/resources/the-social-economy-a-means-for-inclusive -decent-work-in-the-circular-economy.

Calisto Friant, Martin, Walter J.V. Vermeulen, and Roberta Salomone. 2020. "A Typology of Circular Economy Discourses: Navigating the Diverse Visions of a Contested Paradigm." *Resources, Conservation and Recycling* 161 (April): 104917. https://doi .org/10.1016/j.resconrec.2020.104917.

Campbell-Johnston, Kieran, Joey ten Cate, Maja Elfering-Petrovic, and Joyeeta Gupta. 2019. "City Level Circular Transitions: Barriers and Limits in Amsterdam, Utrecht and The Hague." *Journal of Cleaner Production* 235: 1232–39. https://doi.org/10.1016/j .jclepro.2019.06.106.

Caplan, Pat. 2017. "Win-Win?." *Food Poverty, Food Aid and Food Surplus in the UK Today* 33 (3): 1–6.

Caraher, M., and E. Dowler. 2014. "Food for Poorer People: Conventional and 'Alternative' Transgressions." In: M. Goodman & C. Sage (Eds.), Food Transgressions: Making Sense of Contemporary Food Politics. (pp. 227-246). Farnham, Surrey: Ashgate. ISBN 9780754679707

Caraher, Martin, and Sinéad Furey. 2017. "Is It Appropriate to Use Surplus Food to Feed People in Hunger? Short-Term Band-Aid to More Deep-Rooted Problems of Poverty." *Food Research Collaboration* 258 (2): 726–42. https://doi.org/http://dx.doi.org/10 .1016/j.ejor.2016.08.065.

Chaturvedi, Ashish, Jai Kumar Gaurav, and Pragya Gupta. 2017. "Circular Economy The Many Circuits of a Circular Economy." STEPS Working Paper 94, Brighton: STEPS Centre

Circle Economy, Fabric TNO, and Gemeente Amsterdam. 2016. "Circular Amsterdam - A Vision and Action Agenmda for the City and Metropolitan Area." 1–47. https://www .amsterdam.nl/bestuur-organisatie/organisatie/ruimte-economie/ruimte-duurzaamheid/ making-amsterdam/circular-economy/report-circular/.

Circular Glasgow. 2016. "Circular Glasgow: A Vision and Action Plan for the City of Glasgow." 31.

Cohen, Boyd, and Pablo Muñoz. 2016. "Sharing Cities and Sustainable Consumption and Production: Towards an Integrated Framework." *Journal of Cleaner Production* 134: 87–97. https://doi.org/10.1016/j.jclepro.2015.07.133.

Curtis, Fred. 2003. "Eco-Localism and Sustainability." *Ecological Economics* 46 (1): 83– 102. https://doi.org/10.1016/S0921-8009(03)00102-2.

Delind, Laura B. 2006. "Of Bodies, Place, and Culture: Re-Situating Local Food." *Journal of Agricultural and Environmental Ethics* 19: 121–46. https://doi.org/10.1007/s10806 -005-1803-z.

Dubbeling, Marielle, Camelia Bucatariu, Guido Santini, Carmen Vogt, and Katrin Eisenbeiß. 2016. *City Region Food Systems and Food Waste Management: Linking Urban and Rural Areas for Sustainable and Resilient Development*. Bonn and Eschborn: Deutsche Gesellschaft für Internationale Zusammenarbeit (GIZ) GmbH www.fao.org/ publications%0Ahttp://www.fao.org/3/a-i6233e.pdf.

Ehnert, Franziska, Niki Frantzeskaki, Jake Barnes, Sara Borgström, Leen Gorissen, Florian Kern, Logan Strenchock, and Markus Egermann. 2018. "The Acceleration of Urban Sustainability Transitions: A Comparison of Brighton, Budapest, Dresden, Genk, and Stockholm." *Sustainability* 10 (3): 1–25. https://doi.org/10.3390/ su10030612.

Ellen McArthur Foundation. 2015. "Growth within: A Circular Economy Vision for a Competitive Europe." *Ellen MacArthur Foundation*, 100. https://www.ellenmacart hurfoundation.org/assets/downloads/publications/EllenMacArthurFoundation_Growth -Within_July15.pdf.

Ellen MacArthur Foundation. 2019. "Cities and Circular Economy for Food." *Ellen Macarthur Foundation*, 66.

FAO. 2018. "Guiding the Transition to Sustainable Food and Agricultural Systems the 10 Elements of Agroecology."

Fassio, Franco, and Bianca Minotti. 2019. "Circular Economy for Food Policy: The Case of the RePoPP Project in the City of Turin (Italy)." *Sustainability* 11 (21): 1–17. https://doi.org/10.3390/su11216078.

Fassio, Franco, and Nadia Tecco. 2019. "Circular Economy for Food: A Systemic Interpretation of 40 Case Histories in the Food System in Their Relationships with SDGs." *Systems* 7 (3): 43. https://doi.org/10.3390/systems7030043.

Food Matters. 2011. *Brighton & Hove CSA : Interim Report*, September.

Food Matters, and BHCC. 2011. "PAN 06 Food Growing and Development: Brighton and Hove City Council's Local Development Framework." 1–26.

Fratini, Chiara Farné, Susse Georg, and Michael Søgaard Jørgensen. 2019. "Exploring Circular Economy Imaginaries in European Cities: A Research Agenda for the Governance of Urban Sustainability Transitions." *Journal of Cleaner Production* 228: 974–89. https://doi.org/10.1016/j.jclepro.2019.04.193.

Geissdoerfer, Martin, Paulo Savaget, Nancy M.P. Bocken, and Erik Jan Hultink. 2017. "The Circular Economy: A New Sustainability Paradigm?" *Journal of Cleaner Production* 143: 757–68. https://doi.org/10.1016/j.jclepro.2016.12.048.

Gernert, Maria, Hamid El Bilali, and Carola Strassner. 2018. "Grassroots Initiatives as Sustainability Transition Pioneers: Implications and Lessons for Urban Food Systems." *Urban Science* 2 (1): 23. https://doi.org/10.3390/urbansci2010023.

Gille, Zsuzsa. 2012. "From Risk To Waste: Global Food Waste Regimes." *Sociological Review* 60 (SUPPL.2): 27–46. https://doi.org/10.1111/1467-954X.12036.

Girardet, Herbert. 2017. "Regenerative Cities." In: Michalos A.C. (eds) Encyclopedia of Quality of Life and Well-Being Research. Springer, Dordrecht. https://doi.org/10.1007/978-94-007-0753-5_103463

Gomiero, Tiziano. 2018. "Agriculture and Degrowth: State of the Art and Assessment of Organic and Biotech-Based Agriculture from a Degrowth Perspective." *Journal of Cleaner Production* 197: 1823–39. https://doi.org/10.1016/j.jclepro.2017.03.237.

Gorenflo, Neal. 2017. *Activating the Urban Commons. Communities.* https://search.proquest.com/docview/1970209827?accountid=17242.

Gravagnuolo, Antonia, Mariarosaria Angrisano, and Luigi Fusco Girard. 2019. "Circular Economy Strategies in Eight Historic Port Cities: Criteria and Indicators towards a Circular City Assessment Framework." *Sustainability* 11 (13). https://doi.org/10.3390/su11133512.

Hebinck, Aniek, Francesca Galli, Sabrina Arcuri, Brídín Carroll, Deirdre O'Connor, and Henk Oostindie. 2018. "Capturing Change in European Food Assistance Practices: A Transformative Social Innovation Perspective." *Local Environment* 23 (4): 398–413. https://doi.org/10.1080/13549839.2017.1423046.

Hinrichs, C Clare. 2003. "The Practice and Politics of Food System Localization" *Journal of Rural Studies* 19: 33–45.

HLPE. 2019. "Agroecological and Other Innovative Approaches for Sustainable Agriculture and Food Systems That Enhance Food Security and Nutrition Summary and Recommendations." October: 1–13.

Hobson, Kersty, and Nicholas Lynch. 2016. "Diversifying and De-Growing the Circular Economy: Radical Social Transformation in a Resource-Scarce World." *Futures* 82: 15–25. https://doi.org/10.1016/j.futures.2016.05.012.

Ilieva, Rositsa T. 2017. "Urban Food Systems Strategies: A Promising Tool for Implementing the SDGs in Practice." *Sustainability* 9 (10): 1707. https://doi.org/10.3390/su9101707.

IPES-Food. 2017. "What Makes Urban Food Policy Happen? Insights from Five Case Studies." 1–112.

IPES Food. 2019. "Towards a Common Food Policy for the European Union." 1–110. http://www.ipes-food.org/_img/upload/files/CFP_FullReport.pdf.

Jehlička, Petr, and Petr Daněk. 2017. "Rendering the Actually Existing Sharing Economy Visible: Home-Grown Food and the Pleasure of Sharing." *Sociologia Ruralis* 57 (3): 274–96. https://doi.org/10.1111/soru.12160.

Jesus, Ana de, and Sandro Mendonça. 2018. "Lost in Transition? Drivers and Barriers in the Eco-Innovation Road to the Circular Economy." *Ecological Economics* 145 (December 2016): 75–89. https://doi.org/10.1016/j.ecolecon.2017.08.001.

Johansson, N., and M. Henriksson. 2020. "Circular Economy Running in Circles? A Discourse Analysis of Shifts in Ideas of Circularity in Swedish Environmental Policy." *Sustainable Production and Consumption* 23: 148–56. https://doi.org/10.1016/j.spc .2020.05.005.

Jones, Andy, Michel Pimbert, and Janice Jiggins. 2011. "Virtuous Circles: Values, Systems and Sustainability." London: IIED and IUCN CEESP.

Jones, Peter, and Daphne Comfort. 2018. "Winning Hearts and Minds: A Commentary on Circular Cities." *Journal of Public Affairs* 18 (4): 1–7. https://doi.org/10.1002/pa .1726.

Jong, Martin De, Simon Joss, Daan Schraven, Changjie Zhan, and Margot Weijnen. 2015. "Sustainable-Smart-Resilient-Low Carbon-Eco-Knowledge Cities; Making Sense of a Multitude of Concepts Promoting Sustainable Urbanization." *Journal of Cleaner Production* 109: 25–38. https://doi.org/10.1016/j.jclepro.2015.02.004.

Jurgilevich, Alexandra, Traci Birge, Johanna Kentala-Lehtonen, Kaisa Korhonen-Kurki, Janna Pietikäinen, Laura Saikku, and Hanna Schösler. 2016. "Transition towards Circular Economy in the Food System." *Sustainability* 8 (1): 1–14. https://doi.org/10 .3390/su8010069.

Keblowski, Wojciech, Deborah Lambert, and David Bassens. 2020. "Circular Economy and the City: An Urban Political Economy Agenda." *Culture and Organization* 26 (2): 142–58. https://doi.org/10.1080/14759551.2020.1718148.

Kirchherr, Julian, Denise Reike, and Marko Hekkert. 2017. "Conceptualizing the Circular Economy: An Analysis of 114 Definitions." *Resources, Conservation and Recycling* 127 (September): 221–32. https://doi.org/10.1016/j.resconrec.2017.09.005.

Kirwan, James, Brian Ilbery, Damian Maye, and Carey Joy. 2014. "The Local Food Programme: Final Evaluation Report." 1–75. http://www.localfoodgrants.org/public/ evaluation_full_report.pdf.

Kiss, Konrád, Csaba Ruszkai, and Katalin Takács-György. 2019. "Examination of Short Supply Chains Based on Circular Economy and Sustainability Aspects." *Resources*, no. 4: 1–21. https://doi.org/10.3390/resources8040161.

Lazarevic, David, and Helena Valve. 2017. "Narrating Expectations for the Circular Economy: Towards a Common and Contested European Transition." *Energy Research and Social Science* 31 (May): 60–69. https://doi.org/10.1016/j.erss.2017.05.006.

Longhurst, Noel, Flor Avelino, Julia Wittmayer, Paul Weaver, Adina Dumitru, Sabine Hielscher, Carla Cipolla, Rita Afonso, Iris Kunze, and Morten Elle. 2016. "Experimenting with Alternative Economies: Four Emergent Counter-Narratives of Urban Economic Development." *Current Opinion in Environmental Sustainability* 22 (June 2016): 69–74. https://doi.org/10.1016/j.cosust.2017.04.006.

Mairie de Paris. 2017. "White Paper on the Circular Economy of Greater Paris." https://doi .org/10.1080/03772063.1963.11486473.

Maye, Damian. 2016. "Smart Cities Food Governance: Critical Perspectives From Innovation Theory And Urban Food System Planning." In M. Deakin, D. Diamantini and N Borrelli. (eds) The Governance of City Food Systems. Fondazione Giangiacomo Feltrinelli, Rome, et al, pp.: 49-67.

———. 2019. "'smart Food City': Conceptual Relations between Smart City Planning, Urban Food Systems and Innovation Theory." *City, Culture and Society* 16 (October 2017): 18–24. https://doi.org/10.1016/j.ccs.2017.12.001.

Meadowcroft, James. 2009. "What about the Politics? Sustainable Development, Transition Management, and Long Term Energy Transitions." *Policy Sciences* 42: 323–40. https://doi.org/10.1007/s11077-009-9097-z.

Messner, Rudolf, Carol Richards, and Hope Johnson. 2020. "The 'Prevention Paradox': Food Waste Prevention and the Quandary of Systemic Surplus Production." *Agriculture and Human Values* 37: 805–817. https://doi.org/10.1007/s10460-019-10014-7.

Mirosa, Miranda, Louise Mainvil, Hayley Horne, and Ella Mangan-Walker. 2016. "The Social Value of Rescuing Food, Nourishing Communities." *British Food Journal* 118 (12): 3044–58. https://doi.org/10.1108/BFJ-04-2016-0149.

Moragues, A., Morgan, K.; Moschitz, H.; Neimane, I.; Nilsson, H.; Pinto, M.; Rohracher, H.; Ruiz, R.; Thuswald, M.; Tisenkopfs, T. and Halliday. 2013. "Urban Food Strategies. The Rough Guide to Sustainable Food Systems." Document developed in the framework of the FP7 project FOODLINKS (GA No. 265287)

Moragues-faus, Ana, and Kevin Morgan. 2015. "Reframing the Foodscape : The Emergent World of Urban Food Policy" *Environment and Planning A: Economy and Space* 47: 1558–73. https://doi.org/10.1177/0308518X15595754.

Moreau, Vincent, Marlyne Sahakian, Pascal van Griethuysen, and François Vuille. 2017. "Coming Full Circle: Why Social and Institutional Dimensions Matter for the Circular Economy." *Journal of Industrial Ecology* 21 (3): 497–506. https://doi.org/10.1111/jiec.12598.

Morgan, Kevin. 2015. "Nourishing the City: The Rise of the Urban Food Question in the Global North." *Urban Studies* 52 (8): 1379–94. https://doi.org/10.1177/0042098014534902.

Morgan, Kevin, and Roberta Sonnino. 2010. "The Urban Foodscape: World Cities and the New Food Equation." *Cambridge Journal of Regions, Economy and Society* 3 (2): 209–24. https://doi.org/10.1093/cjres/rsq007.

Mourad, Marie. 2016. "Recycling, Recovering and Preventing 'Food Waste': Competing Solutions for Food Systems Sustainability in the United States and France." *Journal of Cleaner Production* 126: 461–77. https://doi.org/10.1016/j.jclepro.2016.03.084.

Neil Ravenscroft, Niamh Moore, Andrew Church Welch. 2012. "Connecting Communities Through Food : The Theoretical Foundations of Community Supported Agriculture in the UK" 44 (115).CRESC Working Paper No. 115

Nogueira, André, Weslynne S. Ashton, and Carlos Teixeira. 2019. "Expanding Perceptions of the Circular Economy through Design: Eight Capitals as Innovation Lenses." *Resources, Conservation and Recycling* 149 (November 2018): 566–76. https://doi.org/10.1016/j.resconrec.2019.06.021.

Olsson, E. Gunilla Almered. 2018. "Urban Food Systems as Vehicles for Sustainability Transitions." *Bulletin of Geography* 40 (40): 133–44. https://doi.org/10.2478/bog-2018-0019.

Ostrom, Marcia Ruth. 2007. "Community Supported Agriculture as an Agent of Change Is It Working ?." In "Remaking the North American Food System", Clare Hinrichs and Tom Lyson (eds). University of Nebraska Press, pp 99-120.

Paiho, Satu, Elina Mäki, Nina Wessberg, Martta Paavola, Pekka Tuominen, Maria Antikainen, Jouko Heikkilä, Carmen Antuña Rozado, and Nusrat Jung. 2020. "Towards

Circular Cities—Conceptualizing Core Aspects." *Sustainable Cities and Society* 59 (January): 102143. https://doi.org/10.1016/j.scs.2020.102143.

Papargyropoulou, Effie, Rodrigo Lozano, Julia K. Steinberger, Nigel Wright, and Zaini Bin Ujang. 2014. "The Food Waste Hierarchy as a Framework for the Management of Food Surplus and Food Waste." *Journal of Cleaner Production* 76: 106–15. https://doi .org/10.1016/j.jclepro.2014.04.020.

Prendeville, Sharon, Emma Cherim, and Nancy Bocken. 2018. "Circular Cities: Mapping Six Cities in Transition." *Environmental Innovation and Societal Transitions* 26: 171– 94. https://doi.org/10.1016/j.eist.2017.03.002.

Preston, Felix. 2012. "A Global Redesign? Shaping the Circular Economy." *Energy, Environment and Resource Governance*, no. March: 1–20. https://doi.org/10.1080 /0034676042000253936.

Rob Hopkins. 2008. "The Transition Handbook. From Oil Dependency to Local Resilience" 3 (2): 54–67. http://repositorio.unan.edu.ni/2986/1/5624.pdf.

Rockström, Johan, Will Steffen, Kevin Noone, and Asa Persson. 2009. "Planetary Boundaries : Exploring the Safe Operating Space for Humanity."

Rockström, J. and Pavan Sukhdev. 2016. "How Food Connects all the SDGs." Stockholm Resilience Centre, June 06, 2016. https://stockholmresilience.org/research/research -news/2016-06-14-how-food-connects-all-the-sdgs.html.

Sánchez Levoso, Ana, Carles M. Gasol, Julia Martínez-Blanco, Xavier Gabarell Durany, Martin Lehmann, and Ramon Farreny Gaya. 2020. "Methodological Framework for the Implementation of Circular Economy in Urban Systems." *Journal of Cleaner Production* 248. https://doi.org/10.1016/j.jclepro.2019.119227.

Sarkis, Joseph, Maurie J Cohen, Paul Dewick, and Patrick Schröder. 2020. "A Brave New World: Lessons from the COVID-19 Pandemic for Transitioning to Sustainable Supply and Production." Resources, Conservation and Recycling 159: 104894

Schröder, Patrick, Alexandre Lemille, and Peter Desmond. 2020. "Making the Circular Economy Work for Human Development." *Resources, Conservation and Recycling* 156 (November 2018): 104686. https://doi.org/10.1016/j.resconrec.2020.104686.

Schroeder, Patrick, Kartika Anggraeni, and Uwe Weber. 2018. "The Relevance of Circular Economy Practices to the Sustainable Development Goals." *Journal of Industrial Ecology* 23 (1): 77–95. https://doi.org/10.1111/jiec.12732.

Schulz, Christian, Rannveig Edda Hjaltadóttir, and P. Hild. 2019. "Practising Circles: Studying Institutional Change and Circular Economy Practices." *Journal of Cleaner Production* 237 (November): 117749. https://doi.org/10.1016/j.jclepro .2019.117749.

Sodiq, Ahmed, Ahmer A.B. Baloch, Shoukat Alim Khan, Nurettin Sezer, Seif Mahmoud, Mohamoud Jama, and Ali Abdelaal. 2019. "Towards Modern Sustainable Cities: Review of Sustainability Principles and Trends." *Journal of Cleaner Production* 227: 972–1001. https://doi.org/10.1016/j.jclepro.2019.04.106.

Sonnino, Roberta. 2010. "Escaping the Local Trap: Insights on Re-Localization from School Food Reform." *Journal of Environmental Policy and Planning* 12 (1): 23–40. https://doi.org/10.1080/15239080903220120.

———. 2016. "The New Geography of Food Security: Exploring the Potential of Urban Food Strategies." *Geographical Journal* 182 (2): 190–200. https://doi.org/10.1111/geoj .12129.

———. 2019. "The Cultural Dynamics of Urban Food Governance." *City, Culture and Society* 16 (January 2017): 12–17. https://doi.org/10.1016/j.ccs.2017.11.001.

Sonnino, Roberta, Cinzia L.S. Tegoni, and Anja De Cunto. 2019. "The Challenge of Systemic Food Change: Insights from Cities." *Cities* 85 (April 2018): 110–16. https://doi.org/10.1016/j.cities.2018.08.008.

Springmann, Marco, Michael Clark, Daniel Mason-D'Croz, Keith Wiebe, Benjamin Leon Bodirsky, Luis Lassaletta, Wim de Vries, et al. 2018. "Options for Keeping the Food System within Environmental Limits." *Nature* 562 (7728): 519–25. https://doi.org/10.1038/s41586-018-0594-0.

Starr-Keddle, H. 2019. "Redistributing Surplus Food in the City has a Big Impact on Carbon Emissions." Brighton and Hove Food Partnership, June 26, 2019. https://bhfood.org.uk/redistributing-surplus-food-in-the-city-has-a-big-impact-on-carbon-emissions/

Temesgen, Amsale, Vivi Storsletten, and Ove Jakobsen. 2019. "Circular Economy: Reducing Symptoms or Radical Change?" *Philosophy of Management* 20: 37-56. https://doi.org/10.1007/s40926-019-00112-1.

Tikka, Ville. 2019. "Charitable Food Aid in Finland: From a Social Issue to an Environmental Solution." *Agriculture and Human Values* 36 (2): 341–52. https://doi.org/10.1007/s10460-019-09916-3.

UNRISD. 2016. "Policy Innovations for Transformative Change. Implementing the 2030 Agenda for Sustainable Development." *Policy Innovations for Transformative Change.* https://doi.org/10.18356/ce8234ef-en.

Velenturf, Anne P.M., and Phil Purnell. 2021. "Principles for a Sustainable Circular Economy." *Sustainable Production and Consumption* 27: 1437–57. https://doi.org/10.1016/j.spc.2021.02.018.

Westley, Frances, Per Olsson, Carl Folke, Thomas Homer-dixon, Harrie Vredenburg, Derk Loorbach, John Thompson, et al. 2011. "Tipping Toward Sustainability : Emerging Pathways of Transformation." AMBIO 40: 762–80. https://doi.org/10.1007/s13280-011-0186-9.

Willett, Walter, Johan Rockström, Brent Loken, Marco Springmann, Tim Lang, Sonja Vermeulen, Tara Garnett, et al. 2019. "The Lancet Commissions Food in the Anthropocene : The EAT – Lancet Commission on Healthy Diets from Sustainable Food Systems" *The Lancet Commissions* 393 (10170): 447-492. https://doi.org/10.1016/S0140-6736(18)31788-4.

Williams, Joanna. 2019. "Circular Cities." *Urban Studies* 56 (November 2017): 2746–62. https://doi.org/10.1177/0042098018806133.

Wiskerke, Johannes S C. 2015. "Urban Food Systems." *Understanding Urban Ecology*, no. September 2015: 307–20. https://doi.org/10.1007/978-3-030-11259-2_14.

WRAP. 2020. "Food Surplus and Waste in the UK: Key Facts." no. January: 14. http://www.wrap.org.uk/sites/files/wrap/Food Surplus and Waste in the UK Key Facts %2822 7 19%29_0.pdf%0Ahttp://www.wrap.org.uk/sites/files/wrap/Food-Surplus-and-Waste-UK-Key-Facts-23-11-18.pdf.

Ziegler, Rafael. 2019. "Water Innovation for a Circular Economy: The Contribution of Grassroots Actors." *Water Alternatives* 12 (2): 774–87.

6 Human waste

Why what seems naturally circular frequently is not

Iris Borowy

Introduction

Given their biological nature one would expect that human excreta lend themselves to a high level of integration into a circular economy (CE) system. In contrast to many manufactured products, they are completely bio-degradable, there is a clear purpose for which they can be put to use, and, in fact, there is a long history of such circular use, in some places going back thousands of years. Considering this background, it is surprising that human waste should not be at the forefront of models of successful CE systems. However, it is not. Firm data about the recycling rate of human waste are virtually non-existent and the practice certainly varies widely between different regions. According to one author, "[g]lobal estimates of current recirculation rates are highly variable, but suggest that, at most, 15 percent of nitrogen and 55 percent of phosphorus in human excreta are recirculated to cropland" (Harder, Wielemaker, Molander, and Oberg 2020: 2). This is considerably worse than some industrial materials. To use an often-cited example, over 60% of aluminium is recycled globally, reaching 74.5% in Europe (Aluminium Association 2014). Other metals reach similar results. A recent UNEP study finds that the end-of-life recycling rate of 18 metals is over 50%, sometimes substantially higher (UNEP 2011). Globally, 35% of glass is recycled (Harder 2018). EU countries recycle a whopping 72% of their paper (Anonymous 2020). Why is it that human waste, despite its excellent preconditions for a circular economy system, does considerably worse than these industrially produced materials? Why is this most organic, most atavistic of human products not the centrepiece of human-made circular transformations, which, at the outset, it should be? In many ways, human waste just formed part of the overall social transformations that came with nineteenth-century industrialisation and fundamentally changed the way people produced, consumed, and discarded.

If circular economy means re-integrating discarded products into production and use, then the term may be new but the practice is certainly not. Before industrialisation, "almost everything was made from materials that were either decomposable – like wood, reeds, or hemp – or easy to recycle or re-use – like iron and bricks" (De Decker 2018). Reusing them was not called "circular economy" then. It was the natural way of living with the normal scarcity of

DOI: 10.4324/9781003255246-6

resources and products. This situation changed with industrialisation and emerging mass production, income, and consumption. With large quantities of cheap new products available, both the need and the possibility to reuse and recycle drastically declined. Over the last two or three centuries, this linear form of economic management became so dominant that, in 2018, Lian Hou, in his preface to *Unmaking Waste*, called it the "traditional linear economy" even though in historical terms it is anything but traditional (Hou 2018: xiii). Instead, industrialisation and the increase in income and wealth it brought created the preconditions to think of a circular economy as environmental progress rather than poverty. Inevitably, the term emerged in a high-income country. Kenneth Boulding is credited with being the first author who coined the expression by writing that "man must find his place in a cyclical ecological system" (Boulding 1966: 3). At the time of publication in 1966, supposedly large part of the world population was still living in predominantly pre-industrial settings and was practising largely circular lifestyles. But Boulding was writing for societies of high-income countries where the challenge of circularity was clearly understood to be one of industrial production and design. Tellingly, one of the formative texts of CE, the 2002 book *Cradle to Cradle* by Michael Braungart and William McDonough, was subtitled *Remaking the Way We Make Things* (Braungart and McDonough 2002). The main thrust of this book and the resulting initiative was towards redesigning production, and, in that vein, some researchers have explored the business opportunities inherent in the CE (Lopes de Sousa Jabbour, Chiappetta Jabbour, Godinho Filho, and Roubaud 2018) Thereby, CE shifted from being normal to being a function of poverty to being a marker of modernity and innovation.

At first sight, human waste seems to have nothing to do with this development. Human excreta are not industrially produced. They cannot be specifically designed, and they do not persist in the environment for a long time the way manufactured materials such as metals or plastics do. Besides, in many cultures, human waste is considered something intensely private and biological, not readily discussed as part of economic policies. And yet, their position in the socio-economic order of almost all societies has seen fundamental changes during the last 150 years, similar to those seen in non-biological forms of production.

In fact, human excreta both confirm and defy the dynamics that determine CE: they are both the quintessential material of circularity, which ties humans to soils, plants, food, and metabolic waste in an endless cycle of organic transformations, and they are the material that disturbs the order of urban living; and they are also both a manifestation of the fundamental equality of humans around the world, all of whom have to eat and excrete to remain alive, and they are a marker of social hierarchies, as connections to human waste has been tied to low social status. Most importantly, their integration into CE has been shaped by much the same considerations of cultural acceptability and social frameworks as industrial products. This paper explores how the dynamics of industrialisation, modernisation, and inclusion or exclusions have been borne out in the context of human waste and its shifting connection to CE.

The principle of human waste and CE

Several points highlight how cultural and socio-economic considerations have determined how or whether human excreta have been organised as CE, much like any other material. To begin with, in industrial production, a turn to CE has been tied to resource scarcity (Hou 2018: xiv). This is no different for human excreta. They were widely used as agricultural fertiliser at a time when fertiliser was scarce since farmers had few other possibilities to replenish fields with nitrogen (animal waste and legumes, whose roots harbour nitrogen-fixing bacteria, being the other important ones). Often, this practice was essential, since access to nitrogen as a crucial building block for proteins – and therefore of the formation of plants and animals – represented a key limiting factor in agricultural productivity. It was the technical access to nitrogen through the Haber-Bosch process, which put an end to this scarcity in industrialised parts of the world and thereby removed a powerful incentive to use human excreta as fertiliser in those farming communities that could afford such industrially produced fertiliser (Smil 2004). In the process, it also connected a CE cycle involving human waste to areas of comparative poverty. Second, while the recycling of human excreta may not depend on the design of this product itself, the design of sanitation systems clearly plays a crucial role. Different designs determine what problems arise for keeping up a circular system or whether it works at all. Third, similar to industrial waste such as discarded oil or plastics, human waste can be hazardous to human health, if not usefully repurposed by re-integration in a circular system or improperly disposed of. Human excreta contain potentially significant amounts of pathogenic viruses, bacteria, cysts of protozoa, and eggs of helminths, which can cause illness and death to people (Singh Maurya 2012: 325–332, 326–329). Fourth, much like industrial production, human waste is produced by humans. It connects human lives with non-human nature by taking up resources and returning discarded material. Just like in industrial production, this process has been commodified, whereby both the raw materials (food) and the disposal comes with costs and profits, creating business opportunities. Fifth, and perhaps most important, just like in industrial production, a circular economy involving human waste requires tying its production, use, discarding, and reproduction into an unbroken circle of constant transformation (Figure 6.1).

To varying degrees, every one of these steps involves negotiations between evolving socio-economic conditions and cultural norms.

Excreta usage (for agriculture)

Human waste comes in different forms: human faeces and urine (i.e., the solid and liquid components of excreta), night soil (a mixture of faeces and urine), sewage (a mixture of water, excreta, and other household and sometimes industrial wastes), and sewage sludge (a solid or semi-solid mass gained from sewage through sedimentation) either in raw form (i.e., right after sedimentation before further transformations) or in digested form, also called biosolids (after transformation

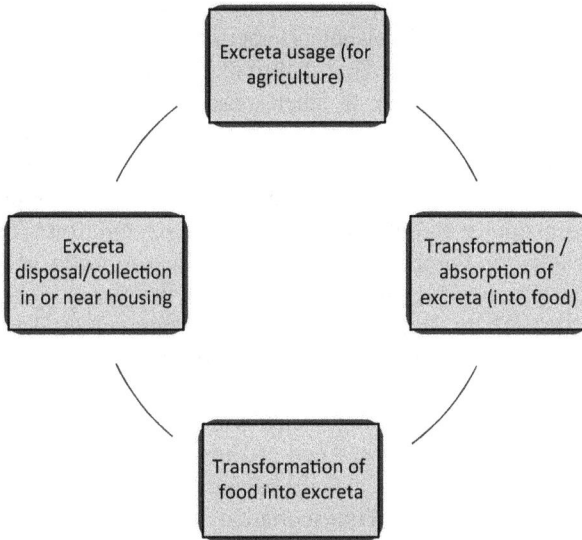

Figure 6.1 The cycle of a human waste CE. Illustration by the author.

through anaerobic or aerobic bacteria). The simplest truth regarding human waste management is that "there is no option to not have significant volumes of excreta and/or wastewater sludge to manage. It exists" (UN-Habitat 2008: 49).

For the vast majority of human history, people have dealt with their waste the way many other animals do. As small nomadic groups moved around, they urinated and defecated out in the open and moved on. The results just formed part of the ongoing circle of organic matter, effectively serving as food for various insects and microorganisms. The question only became problematic after humans gave up their nomadic lifestyles and took up agriculture. Like animals who live in caves or build nests, humans now had to see to it that they relieved themselves in ways that avoided fouling the places where they slept, played, and ate. Inevitably, humans lived in closer proximity to their excreta than before, and a higher load of pathogens, i.e., a higher burden of faecal-borne diseases, was one price to be paid for the benefits of sedentary living. Typhoid fever and schistosomiasis are therefore among the first diseases that afflicted humans after they became sedentary (Humphreys 1997: 14–19; Farley 1997: 20–23). In the process, increases in social life (as more people could connect to form large social groups) and in culture (as more members of these social groups could focus on painting, writing, or composing instead of growing food) came at the price of more intense connection to excreta. Instead of just returning organic material into nature wherever this nature had called, human waste now became material, which farmers could deliberately employ as part of the cultivation of their land in order to fertilise the soil and improve the yields of their harvests.

Not all societies have used human excreta. Where land was plentiful, farmers often used swidden agriculture or left fields lay fallow to allow the soil to replenish itself without engaging in elaborate practices of soil enhancement. In the United States, these methods were used until the late nineteenth century (Rogers 2005: 33–34; Federico 2005: 8). In other areas, access to alternatives, notably from domestic animals, reduced the need for human excrement as fertiliser (Brunt 2007: 333–372). Farmers in still other areas, notably regions in Sub-Saharan Africa, New Zealand, and Polynesia, did not make use of human excrement for reasons that seem not yet entirely clear (Webb 2019: 45–49). But in many parts of the world, including in Europe, Africa, pre-Columbian America, and, especially, in Asia recycling human excrements was common practice (Leigh 2004; Black and Fawcett 2008: 27–29; Federico 2005: 8–9). As farmers recycled their excrements into the soil from which they drew their food, they practised a simple local circular economy.

The situation became more complicated with the formation of cities, which physically separated the metabolised end product of plant food (human excreta) from the places where it could become the beginning of new plant food (farming), so that closing this cycle required transportation.

Arguably, the country that made the most intensive use of night soil was Japan, where a growing population and few domestic animals made human waste extremely valuable in the modern period. By the eighteenth century, villagers were in active competition for the right to collect the excrement of specific urban districts, considered to have different agricultural and economic value (Hanley 1987). Similarly, from the late Ming Dynasty onwards, many Chinese cities, especially in the south, had an elaborate system of trade in night soil, graded into various quality categories, according to the diet of their producers. Some wealthy cities, such as Hangzhou and Souzhou, entertained lively markets in high-quality night soil, maintained with the help of professional night soil collectors, which made it worthwhile for farmers to travel long distances to buy excrement (Masuki 2018; Xue 2005; Marks 2012: 166–167). Using excrement as fertiliser also has had a long tradition in Flanders. Much like for cities in China,

> Brussels' faeces had a market value and constituted a substantial source of income for the municipal administration. As such, it indirectly contributed to the wealth of the urban population as much as the urban population contributed to the growth of the yields and profits of agriculture.
>
> (Deligne 2016: 249–250)

For millennia, urban and rural areas were linked through the movement not only of food, goods, and people but also of organic waste. Thus, "in a double- helix-like transformation, farming and urbanization dialectically shaped each other through waste handling" (Rogers 2005: 32). When and where these exchanges of food for night soil functioned, this system worked as a geographical extension of the circular system farmers had had before. Special, sometimes very elaborate arrangements allowed maintaining a circular economy even as production and consumption grew further apart.

However, this system had its limits. The explosive growth of the world population from one billion people in 1800 to around 7.7 billion in 2020 has made this arrangement more difficult (van Bavel 2013). Sprawling urbanisation lengthened the distance between the production of human waste and farms that could use them, making the transportation of human waste to the farms that needed them more inconvenient and expensive. In Europe and North America, the increasing availability of new fertilisers, guano, and, increasingly, artificial fertiliser, provided convenient and increasingly affordable alternatives (McNeill 2000: 22–26). In addition, the adoption of flush toilets and water sanitation in a growing number of industrialising countries and regions around the world proved a game-changer by effectively dividing the world into two parts: one in which human waste was flushed into the sewage and another in which it was not, whereby one was considered hygienic, convenient, and modern while the other was considered unhealthy, troublesome, and backward (Nick Cullather 2004: 230).

This development transformed human excreta from potentially valuable agricultural material to potentially hazardous waste. It also tied sanitation to water management. This identification was not complete, as some people working in the development field did perceive the benefit of alternative methods (Borowy 2021). But it was far-reaching and to the detriment of a holistic thinking of excreta. As a case in point, the Millennium Development Goals (MDGs) of the year 2000 framed excreta disposal as a function of water management by calling for "sustainable access to safe drinking water and basic sanitation" in target 7C (UN undated). Even then, sanitation was only an afterthought, added two years after the original agreement of the goals (Satterthwaite and McGranahan 2006: 9).

No doubt, flushing has been convenient for its users, and it did solve some problems associated with keeping excreta near housing. But it was only available in high-income societies, and it made keeping human waste in circular usage problematic. This element was criticised early on, as observers saw clearly that flushing risked getting rid not only of the nuisance of excreta but also of their nutrient value. Some people regarded this solution as a terrible waste of valuable resources. Chemist Justus von Liebig and agriculturist Alderman Mechi protested. Von Liebig calculated that the flush sewage system in London represented a loss of income worth four million British pounds, while Mechi suggested that a separate system of pipes could directly supply farmers who would only need to open their sewage taps to get fertiliser (George 2008/2009: 175). Similarly, in the early twentieth century, an American visitor to China compared the constructive Chinese system of excreta management with the wasteful American practice of flushing fertiliser into rivers and seas instead of using it to restore soil fertility (Donald Worster 2017: 16.)

A sewage tap never materialised, but for a while, many places reconciled flush toilets with the recycling of excreta. Sewage farms served to adapt this largely liquid form of human excreta to the circular arrangement of cities supplying surrounding rural areas with human waste as fertiliser, receiving food in return.

Such sewage irrigation was practised near urban areas in Australia, India, South Africa, and the USA. Before 1939, both Paris and Berlin operated sewage farms of 52,500 and 11,000 hectares respectively (Ignatieff and Page 1949/1958: 43). However, sewage farms were always problematic. They required a lot of land, which had to absorb a constant supply of water, regardless of whether it was already soaked in rainwater. They also faced increasing problems of acceptability because of odours and rising concerns regarding health hazards. By mid-century, sewage farms were largely given up (Hardy 2005: 98–107, 139–166). However, the principle remained attractive for dry areas in which leaving large amounts of water unused was an unaffordable luxury. A World Bank study in the 1980s found that wastewater reuse in agriculture was still very much alive, in industrial as well as in low-income countries, though the findings differed substantially between countries, ranging from virtually non-existent in Britain, to limited in India, China, and Japan to widespread use in Israel and South Africa (Shuval et al. 1986: 8–17). The practice was not limited to land agriculture but proved easily adaptable to aquafarming, where, by 1989, globally "at least two-thirds of the world yield of farmed fish came from ponds fertilized in this way" (Mara and Cairncross 1989: 3–4).

Another method emerged naturally when the increasing pollution of rivers and other waterways prompted the installation of treatment plants, which separated polluted liquid into clean water and sludge. In the United States, municipalities were not required to treat their sewage until the Clean Water Act of 1972. Sixteen thousand sewage treatment plants sprung up, resulting in cleaner water and more sludge. After dumping the latter off the coasts resulted in toxic shellfish beds and algal blooms, the Ocean Dumping Ban Act prohibited this practice in 1989. At the time, Americans produced seven million dry tonnes of sludge per year (George 2008/2009: 174). The problem with this method was, of course, that this sludge treatment contained not only human excreta but whatever else was discarded through sewage. Indeed, human waste removal was not the first and never the only purpose of sewers. Early sewage pipes mainly served to remove stormwater. Later, industry used them for the disposal of their effluents (Angelakis et al. 2014). Rose George quotes a label which a sewage manager in Utah had placed on bottled treated water, which listed possible ingredients as "water, faecal matter, toilet paper, hair, lint, rancid grease, stomach acid and trace amounts of Pepto, Bismol, chocolate, urine, body oils, dead skin, industrial chemicals, ammonia, soil, laundry soap, bath soap, shaving cream, sweat, saliva, salt, sugar" (George 2008/2009: 169). The list could easily be extended to include microplastics, traces of medical and recreational drugs, wet wipes, food leftovers or cooking oil. In recent years, the common combination of grease and wet wipes or similar material has repeatedly led to the formation of "fatbergs" which regularly clog pipes and sewage in various parts of the world. They are an expensive nuisance. New York City alone spent $18 million on fighting fatbergs between 2011 and 2016 (Engelhaupt 2017). They are also an all too visible sign of the quantity of matter other than human excreta, which finds itself in global sewers.

Predictably, this mixture of material has affected the quality and demand for sludge-based fertiliser. Already in 1966, long before the word "fatberg" had been coined, a survey of 21 composting plants in ten (mostly European) countries found that the compost often contained splinters of glass, metals, and plastics. Given that this product competed with high-quality synthetic fertiliser, demand was low and sales slow. In most plants, producing compost was more expensive than land-filling or incinerating the sludge even when proceeds from sales were factored in (Kupchik 1966; Shuval, Gunnerson, and Julius 1981: 13–15). But apparently, alternatives were difficult to find. By 1981, there were an estimated 200 municipal composting plants in use in Europe (Office of Solid Waste Management Program 1971: 7).

Today, sludge can contain any number of 100,000 chemicals used by industry, endocrine-disrupting PCBs and phthalates, carcinogenic dioxins, or a variety of pathogens. Given this context, it is remarkable that sludge, reframed as "biosol-ids", is still used as fertiliser in any form. In fact, in the United States, it is a profitable business, worth millions of dollars. The material undergoes substantial processing, transforming it into powder or pellet, and it is categorised into dif-ferent levels of purity, to be used for different purposes. Only a high categorisa-tion is meant to be used for growing food (George 2008/2009: 173–175). But the practice is problematic, as health problems in the next phase of the circle of use show. In reality, the mixed and partially toxic nature of sewage sludge means that composting, incinerating, landfilling and dumping have all become increas-ingly problematic, so that industrialised countries are struggling to find a solution within non-circular or circular methods (Hudcova, Vymazal, and Rozkošný 2018; Bianchini, Bonfiglioli, Pellegrini, and Saccani 2016; Ivanova et al. 2018).

The transformation/absorption of excreta (into food)

In principle, the transformation of excreta into compost and thereby, eventually, into food (for humans and other living beings) is a natural process which takes place by itself when human waste is released into natural surroundings and, as mentioned, its employment in agriculture can look back to a long history. Human excreta contain crucial plant nutrients, notably nitrogen, phosphorus, and potas-sium, as well as partially healthful or harmful trace elements like calcium, mag-nesium, zinc, copper, nickel, cadmium, lead, mercury, and barium, which lend themselves to useful re-integration into the production process. Comparatively, urine is richer in nutrients, but faeces are high in material improving soil structure (Sing Maurya 2012: 325–332, 326–329). The arrival of synthetic fertiliser effec-tively ended this age-old connection in industrialised regions. However, its price has kept it out of reach of substantial parts of the world population, so the use of human waste for this purpose never ended completely. In addition, as became increasingly apparent, synthetic fertiliser came with negative social and environ-mental effects (McNeill 2000: 26–30).

Human waste as a fertiliser, therefore, never quite went away. But its use has been substantially curtailed by concerns about its potentially pathogenic character.

Indeed, the disease potential of untreated excreta is impressive. At present, intestinal pathogens infect hundreds of millions of people every year, mainly children. High impact diseases include viral infections such as polio, hepatitis, or norovirus, bacterial infections such as typhoid fever (*Salmonella typhi* and *Salmonella paratyphi*) and cholera (*Vibrio cholerae*), protozoa infections by *Giardia lamblia* and *Entamoeba histolytica* (causing amoebic dysentery), and worm infections such as schistosomiasis or infections with soil-transmitted helminths such as roundworms, whipworms, and hookworms (Webb 2019: 1–6). There are also more mundane infections, which do not kill but cause various degrees of diarrhoea, severely inhibiting the nutritional status, growth, and development of children.

The beginning of germ theory and a growing understanding of pathogens in the nineteenth century led to rising awareness of the potential dangers of human waste, especially in industrialised countries. A growing, increasingly urban population saw human waste not as valuable agricultural input but primarily as carriers of pathogens, developing a "faecal aversion barrier" (Un-Habitat 2008: 4). This attitude both spurred and was strengthened by the installation of flush toilets, which eliminated excreta from people's houses and minds. Even today, preventing death and disease is widely perceived as the big achievement of flush waste disposal. In 2007, 11,300 readers of the *British Medical Journal* voted the introduction of clean water and sewage the most important medical achievement since 1840 (Ferriman 2007). This view has made the idea of connecting human excreta with food seem foolhardy at best and irresponsible at worst. The received scientific wisdom was that it was a risky practice, acceptable only as a measure of desperation for those to whom no other fertiliser was available (Gotaas 1956).

More recent research has begun questioning this perspective. While the connection between disease and human waste is uncontested, the role of its use in farming is far less certain. Already in the 1980s, a large meta-study of epidemiological studies indicated that infections depended on many factors beyond the presence of pathogens, including the immune status of people, climate, the choice of plants, and preparation of food. In the 1980s, empirical data suggested that the health hazards were largely limited to the transmission of various helminths if vegetables irrigated with untreated wastewater had been consumed uncooked, as well as to agricultural workers. Evidence for the transmission of bacterial or viral diseases was thin (Shuval 1986: 27–45, 123–137). Over the following years, further studies seemed to corroborate those findings, and international organisations such as the World Health Organization and the World Bank gradually shifted their recommendations from avoiding the use of human waste as fertiliser, if at all possible, to one embracing it under circumstances that guaranteed safety (Borowy 2021). Increasingly, the threat, which for a long time appeared like the major stumbling block for the integration of human waste into a circular system, seems controllable.

Meanwhile, it is the toxic burden of sludge due to chemical contamination which has become the primary concern. For instance, in the 2000s, families in Virginia living near fields fertilised with sludge have documented clusters of cancer in the neighbourhood (George 2008/2009: 183–185). Opposition has also sprung up among environmentalist groups. The Sierra Club in Washington State uses its website to explain its opposition to "the use of contaminated toxics-containing or pathogen-containing waste as a compost ingredient and the application of municipal sewage sludge as a fertilizer" and to appeal to readers to join its "Sewage Sludge Free Washington" campaign (Sewage Free Washington undated).

Meanwhile, "many countries control not only the concentration of heavy metals in biosolids applied to soils, but also restrict the amounts of particular heavy metals that can be applied within a given period of time"(UN-Habitat 2008: 33).

Clearly, a system that mixes human with other, partly toxic waste is not conducive to a circular economy. If too much of the waste is too much of a health hazard to be transformed into food, it cannot enter the stage of transformation from food into excreta, and the circle is interrupted.

Transformation of food into excreta

The transformation of food into excreta is an inevitable physiological process, and there is not a lot that could or needed to be done to influence this part of the circle. As long as people eat, there will be human waste.

In fact, population growth has made sure that, by any standards, there is a lot of it. Depending on diet, stature, and climate, the average person is believed to produce between 100 and 500 grams of faeces or between 1,000 and 1,700 grams of excreta (faeces and urine) per day (Singh Maurya 2012: 325–332, 326–327). Given a world population of approximately 7.7 billion people, this amounts to between 280 million and 1.4 billion tonnes of faeces or between 2.8 and 4.8 billion tons of excreta on any given day. This means that day after day people, societies, and administrations have to decide, by design or by default, what to do with massive amounts of a material which can, potentially, be fully returned into an unending cycle of use and consumption.

However, how to feed the material from its place of production into this cycle, is not that simple. Regardless of whether flush toilets are used or not, all methods come with the challenge of excreta collection and transportation.

Excreta disposal/collection in or near the home

Due to the potentially pathogenic character of excreta, keeping it in or near the home in ways that leave them accessible to people and disease vectors, notably flies, creates substantial health risks. In many parts of the world, the traditional way to keep human waste away from the house has been to leave it in the open, at some distance from the home. In slum areas with little open space, this sometimes takes the modern variation of "flying toilets", to place faeces into a plastic bag and throw it away (Asabia 2009). At the time of writing, an estimated 673 million people practice open defecation, while two billion people do not have basic sanitation facilities such as toilets or latrines. Poor sanitation is a leading cause of diarrheal diseases causing undernutrition and 432,000 annual deaths (UN News 2019; WHO 2020).

People in industrialised countries solved this problem – or thought they did – by installing flush toilets and establishing an immense network of underground sewage pipes. Several factors stand in the way of universal spread of flush toilets: they require large quantities of water, which is problematic in dry regions, and they place the potential fertiliser out of reach for farmers. Above all, its infrastructure is expensive to build, especially in already existing large urban areas. A 2006 UNDP study lists the price of sanitation options as ranging from zero for open defecation and "flying toilets" to $400–1,500 for flush toilets connected to sewers and piped water. In between, there are communal latrines, starting at $12, and "ecological sanitation" with provisions for nutrient recycling at $90–350 and more (Satterthwaite and McGranaham 2006: 7–8). These price tags mean that "up to 70 per cent of populations in cities in sub-Saharan African countries and in countless towns and cities in South Asia and elsewhere are going to have to rely for the foreseeable future on pit toilets" (Black and Fawcett 2008: 79). In fact, even poor and neglected communities in the USA, the richest country in the world, suffer from lack of access to public sewage and face a fine when they cannot afford expensive private arrangements of septic tanks (Pilkington 2021).

Badly planned sanitation methods are not only harmful to health; they also stand in the way of CE by interrupting the cycle of usage. An obvious prerequisite is that human excreta are collected in a way that they can be made available for agricultural use. This requires that 1) people must use toilets which collect excreta, and 2) excreta must be moved from these places to where they can be re-integrated into the circle. Getting people to use toilets can be problematic not only because of the costs involved. Even people who could afford toilets do not always have them installed. Habit, a cultural belief in the cleanliness of open-air defecation, an unwillingness to invest money in something that has, so far, been free, the unappealing nature of existing toilets because of filth, insects, or lack of privacy can all stand in the way of an easy embrace of toilet usage. Rural social workers can go to great lengths to try to convince communities to invest in sanitation (George 2008/2009: 195–224). Sometimes, measures taken for different reasons can be unexpectedly helpful for this purpose. When Kenya prohibited plastic bags in 2017, it all but eliminated "flying toilets" and increased the use of public toilets (Watts 2018).

As a next step, someone needs to empty the latrines or pit toilets and transport the contents to where they can be recycled into fertiliser material. It is easy to forget that, as recently as the 1960s, well-organised systems of human waste transportation still existed in Japan and Hong Kong (Forbes 1967; Pradt 1971). Today's systems are rarely as well organised. Usually, the job is left to the informal sector, and usually sanitation workers do not enjoy legal protection. Sanitation work carries a social stigma, it is generally badly paid, leaving workers financially insecure with little bargaining power, and it is unhealthy and dangerous, especially since sanitary workers often lack even the most basic protective equipment. The list of common health problems is long, including headaches, dizziness, fever, fatigue, asthma, gastroenteritis, cholera, typhoid, hepatitis, schistosomiasis, eye and skin burn and other skin irritation, puncture wounds and cuts, and blunt force trauma. Noxious gases in excreta containers regularly cause workers to lose consciousness and, in some instances, die. Deaths are also caused by collapsing pits, falling masonry, or wounds received from sharp detritus (World Bank/WHO/ILO/Water Aid 2019: 8–9). A 2019 report, jointly published by the World Bank, WHO, ILO, and Water Aid, calls for the formalisation of sanitation work, including providing legal protection, promoting unionisation or other occupational association, reducing health risks and providing health services, and establishing standard operating procedures and guidelines (World Bank/WHO/ILO/Water Aid 2019: 10). The implementation of such policies can run counter deeply entrenched cultural biases. A case in point is India, where concepts of purity and pollution form an important part of Hinduism and underwrite a caste system where social prejudices in combination with the availability of cheap Dalit labour impede sanitary improvements and have proved difficult to change (Gatade 2015).

Sending human waste into sewers means they end up either in rivers and, ultimately, the sea, or as sludge in treatment plants. If rivers containing human waste are used for irrigation, this is a circularity of sorts, but in an uncontrolled and potentially unsafe manner. Sludge can be and has been used for agricultural fertiliser, though this practice is increasingly considered problematic. Forms involve the difficulties discussed above.

The turn of the century saw a profound change in attitude towards sanitation within international organisations. Publications dedicated to "sanitation" or "sanitary" issues, long niche topics, virtually exploded in the 1990s, reflecting a surge in related research (Zhou 2018). The Sustainable Development Goals (SDGs), though still framing sanitation as part of water disposal, did include a reference to "wastewater treatment, recycling and reuse technologies" in target 6.a (UN DESA undated). In 2017, an influential report by an NGO in India conceptualised sanitation not as mere disposal of human waste, ideally through flushing, but as a comprehensive system consisting of a series of elements: interface, containment/emptying, transport, treatment, and disposal, including re-use in agriculture as manure (Dahlberg Advisers 2017). This concept has since been taken up by other organisations, including the World Bank, whose website explicitly includes on-site as well as sewered solutions and considers resource recovery and re-use. A promotional video explicitly debunks the "myth" that sludge is merely waste to be eliminated (World Bank undated).

Reuse was also included in proposals in 2012 when the Bill and Melinda Gates Foundation announced a design competition for "a revolutionary toilet that was safe, sustainable, and affordable for people in these areas" (Weller 2016). Winning designs worked without access to sewage and usually involved environmental benefits such as electricity production and/or the transformation of the waste into various forms of fertiliser.

Projects along these lines have since continued and may well foreshadow the development of toilets of the future.

Conclusions

The relation between human waste and CE is socially negotiated. The four components circle indicating CE for human waste is certainly a simplification of real-life circumstances, but it serves the purpose of pointing out main areas. Two of those areas are mainly biological and seem to take care of themselves: humans, when eating, produce excreta, and excreta, when placed in the soil, will support the growth of plants that humans will eat. However, these processes also take place within a social and cultural context: the contents of the excreta depend on consumed food, and the degree to which excreta can contribute to growing healthy food depends on the extent to which they contain hazardous pathogens or industrial toxins. But it is predominantly the two areas in between, which are socially engineered and require planning and organisation. For a functioning CE, they also need a suitable conceptual approach and attitude. In that vein, Nityanand Singh Maurya has called for abandoning end-of-pipe technology and embracing a holistic approach instead (Sing Maurya 2012: 326). In practical terms, this might mean giving up flush toilets in favour of on-site collection; separating private from industrial sewage; providing a well-equipped, formal sanitary workforce; or any combination of these and other points. Overall, it is important that the management of excreta is understood as an integral component of food production and should receive similar attention, resources, and prestige. In short, it might best be expressed in what William McDonough, one of the authors of *Cradle to Cradle*, summed up as "Waste Equals Food" (McDonough 1998).

While this analogy is specific to human waste, the underlying principle of thinking circularly instead of linearly seems the same for every section of human material life.

Bibliography

The Aluminium Association. 24 September 2014. "Aluminum can recycling holds at historically high levels." https://www.aluminum.org/news/aluminum-can-recycling-holds-historically-high-levels

Angelakis, Andreas, et al. 2014. "The historical development of sewers worldwide." *Sustainability* 6: 3936–3974.

Anonymous. 2020. "Recycling rate for paper increased to 72%." *Recycling Magazine*, 18 July 2020. https://www.recycling-magazine.com/2020/07/18/recycling-rate-for-paper-increased-to-72/

Asabia, A. 2009. "Presentation on public, private and people partnership model for sanitation at the 6th earth watch conference on water and sanitation 2008." *International NGO Journal* 4 (8): 362–367. Available online at http:// www.academicjournals.org/INGOJ

van Bavel, Jan. 2013. "The world population explosion: causes, backgrounds and projections for the future." *Facts, Views and Vision in Obstetrics and Gynaecology* 5 (4): 281–291.

Bianchini, Augusto, Luca Bonfiglioli, Marco Pellegrini and Cesare Saccani. 2016. "Sewage sludge management in Europe: a critical analysis of data quality." *International Journal of Environment and Waste Management* 18 (3): 226–238.

Black, Maggie and Ben Fawcett. 2008. *The Last Taboo*. Milton Park: Earthscan.

Borowy, Iris. 2021 "Human excreta: hazardous waste or valuable resource? Shifting views of modernity." In Iris Borowy (guest ed.), Special Issue: Development in World History: Development as World History, *Journal of World History* 32 (3) 2021, 517–545. https://muse.jhu.edu/issue/46077.

Boulding, Kenneth E. 1966. "The economics of the coming spaceship earth." In *Environmental Quality in a Growing Economy*, edited by H. Jarrett, 3–14. Baltimore, MD: John Hopkins University Press.

Braungart, Michael and William Mc Donough. 2002. *Cradle to Cradle. Remaking the Way We Make Things*. Berkeley, CA: North Point Press.

Brunt, Liam. 2007. "Where there's muck, there's brass: the market for manure in the industrial revolution." *Economic History Review* 60 (2): 333–372.

Cullather, Nick. 2004. "Miracles of modernization: the green revolution and the apotheosis of technology." *Diplomatic History* 28 (2): 227–254.

Dalberg Advisors. 2017. "Sanitation workers safety and livelihoods in India. A blueprint for action." Phase 4: City Blueprint, Trichy, 6. https://www.susana.org/_resources/documents/default/3-3483-7-1542726225.pdf, retrieved 22 April 2021.

De Decker, Kris. 2018. "How circular is the circular economy?" *Uneven Earth* 27 November (2018), http://unevenearth.org/2018/11/how-circular-is-the-circular-economy/, retrieved 16 March 2021.

Deligne, Chloe. 2016. "Industrialisation, manure and water quality in the 19th century. The Senne River in Brussels as a case study." *Water History* 8: 235–255.

Engelhaupt, Erika. 2017. "Huge blobs of fat and trash are filling the world's sewers." *National Geographic* 17 August 2017, https://www.nationalgeographic.com/science/article/fatbergs-fat-cities-sewers-wet-wipes-science, retrieved 21 April 2021.

Farley, John. 1997. "Snail fever: schistosomiasis, or is it bilharzia?" In *Plague, Pox and Pestilence. Disease in History*, edited by Kenneth Kiple, 20–23. London: George Weidenfeld & Nicolson.

Federico, Giovanni. 2005. *Feeding the World*. Princeton, NJ: Princeton University Press.

Ferriman, Annabel. 2007. "*BMJ* readers choose the 'sanitary revolution' as greatest medical advance since 1840." *British Medical Journal* 334(7585): 111. Doi: 10.1136/bmj.39097.611806.DB.

Forbes, G.I., J.D.F. Lockhart and R.K. Bowman. 1967. "Cholera and nightsoil infection in Hong Kong, 1966." *WHO Bulletin* 36: 367–373.

Gatade, Subhash. 2015. "Silencing caste, sanitising oppression. understanding swachh bharat abhiyan." *Economic & Political Weekly* October 31, 2015, l (44): 29–35.

George, Rose. 2008, paperback edition 2009. *The Big Necessity*. London: Portobello Books.

Gotaas, Harold. 1956. *Composting. Sanitary Disposal and Reclamation of Organic Wastes*. Geneva: WHO.

Hanley, Susan B. 1987. "History urban sanitation in preindustrial Japan." *The Journal of Interdisciplinary History* 18 (1): 1–26.

Harder, Joachim. 2018. "Glass recycling: current market trends." *Recover* May 2018, https://www.recovery-worldwide.com/en/artikel/glass-recycling-current-market-trends_3248774.html.

Harder, Robin, Rosanne Wielemaker, Sverker Molander, and Gunilla Oberg. 2020. "Reframing human excreta management as part of food and farming systems." *Water Research* 175 (115601): 1–8. https://doi.org/10.1016/j.watres.2020.115601.

Hardy, Anne. 2005. *Ärzte, Ingenieure und städtische Gesundheit. Medizinische Theorien in der Hygienebewegung des 19. Jahrhunderts*. Frankfurt/M.: Campus.

Hou, Lian. 2018. "Preface." In *Unmaking Waste in Production and Consumption: Towards the Circular Economy*, edited by Robert Crocker, Christopher Saint, Guanyi Chen and Yindong Tong, Xiii–xix. Bingley: Emerald Publishing.

https://www.who.int/news-room/fact-sheets/detail/millennium-development-goals-%28mdgs%29

Hudcova, Hana, Jan Vymazal and Miloš Rozkošný. 2018. "Present restrictions of sewage sludge application in agriculture within the European Union." *Soil and Water Research* 14 Online: 1–17, https://doi.org/10.17221/36/2018-SWR

Humphreys, Margaret. 1997. "Typhoid and its carriers." In *Plague, Pox and Pestilence. Disease in History*, edited by Kenneth Kiple, 14–19. London: George Weidenfald & Nicolson.

Ignatieff, Vladimir and Harold Page. Revised and enlarged ed. 1958. First ed. 1949. *Efficient Use of Fertilizers*. FAP Agricultural Studies No.43. Rome: FAO.

Ivanova, Lucia et al. 2018. "Pharmaceuticals and illicit drugs: a new threat to the application of sewage sludge in agriculture." *Science of the Total Environment* 634: 606–615.

Kupchik, George. 1966. "Economics of composting municipal refuse in Europe and Israel, with special reference to possibilities in the USA." *WHO Bulletin* 34: 798–809.

Leigh, G.J. 2004. *The World's Greatest Fix: A History of Nitrogen and Agriculture*. Oxford: Oxford University Press.

Lopes de Sousa Jabbour, Ana Beatriz, Charbel Jose Chiappetta Jabbour, Moacir Godinho Filho and David Roubaud. 2018. "Industry 4.0 and the circular economy: a proposed research agenda and original roadmap for sustainable operations." *Annals of Operations Research*, Springer 270 (1): 273–286. DOI: 10.1007/s10479-018-2772-8.

Mara, Duncan and Sandy Cairncross. 1989. *Guidelines for the Safe Use of Wastewater and Excreta in Agriculture and Aquaculture. Measures of Public Health Protection*. Geneva: WHO.

Marks, Robert. 2012. *China. Its Environment and History*. Lanham: Rowman & Littlefield.

Masuki, Yui. 2018. "Historical development of low-cost flush toilets in India: Gandhi, Gandhians, and 'Liberation of Scavengers'." *Sanitation Value Chain* 2 (1): 3–26.

McDonough, William. 1998. "Waste equals food." In *Awakening: The Upside of Y2K*, edited by Judy Laddon, Tom Atlee & Larry Shook. The Printed Word. Available online: https://ratical.org/co-globalize/waste=food.html.

McNeill, John. 2000. *Something New Under the Sun*. New York: W.W. Norton & Co.

Office of Solid Waste Management Program. 1971. *Composting of Municipal Solid Wastes in the United States*. Washington DC: U.S. Environmental Protection Agency.

Pilkington, Ed. 2021. "Interview. Activist Catherine flowers: the poor living amid sewage is 'the final monument of the confederacy'." *The Guardian*, 11 February 2021. https://www.theguardian.com/us-news/2021/feb/11/catherine-flowers-environmental-justice-sewage-alabama

Pradt, Louis. 1971. "Some recent development in night soil treatment." *Water Research* 5: 507–521.

Rogers, Heather. 2005. *Gone tomorrow*. London/New York: The New Press, 2005.

Satterthwaite, David and Gordon McGranahan. 2006. *Overview of the Global Sanitation Problem*. Human Development Report Office Ossasional Paper. 2006/12. Nairobi: UNDP.

Sewage Sludge. n.d. *Free Washington*. https://www.sierraclub.org/washington/north-olympic/sewage-sludge-free-washington, retrieved 21 April 2021.

Shuval, Hillel, Charles Gunnerson and DeAnne Julius. 1981. *Night-soil Composting*. Washington, DC: World Bank.

Shuval, Hillel et al. 1986. *Integrated Resource Recovery. Wastewater Irrigation in Developing Countries*. World Bank Technical Paper 51. Washington, DC: World Bank.

Singh Maurya, Nityanand. 2012. "Is human excreta a waste?" *International Journal of Environmental Technology and Management* 15 (3/4/5/6): 325–332.

Smil, Vaclav. 2004. *Enriching the Earth: Fritz Haber, Carl Bosch, and the Transformation of World*. Cambridge, MA: MIT Press.

UN DESA. n.d. *Goal 6*, https://sdgs.un.org/goals/goal6, retrieved 22 April 2021.

UNEP. 2011. *Recycling Rates of Metals. A Status Report*. Nairobi: UNEP.

UN-Habitat. 2008. *Global Atlas of Excreta, Wastewater Sludge, and Biosolids Management: Moving Forward the Sustainable and Welcome Uses of a Global Resource*. Nairobi: UN-Habitat.

UN News. 2019. "'Transformational benefits' of ending outdoor defecation: why toilets matter." 18 November 2019, https://news.un.org/en/story/2019/11/1051561, retrieved 22 April 2021.

United Nations. n.d. *MDG 7*, https://www.un.org/millenniumgoals/environ.shtml, retrieved 22 April 2021.

Watts, Jonathan. 2018. "Eight months on, is the world's most drastic plastic bag ban working?" *The Guardian*, 25 April 2018. https://www.theguardian.com/world/2018/apr/25/nairobi-clean-up-highs-lows-kenyas-plastic-bag-ban.

Webb, James. 2019. *The Guts of the Matter*. Cambridge: Cambridge University Press.

Weller, Chris. 2016. "4 toilets of the future that are backed by Bill Gates." *Insider*, 16 March 2016, https://www.businessinsider.com/innovative-toilets-backed-by-bill-gates-2016-3, retrieved 26 April 2021.

WHO. 2019. *Factsheet: Sanitation*, 14 June 2019. WHO, https://www.who.int/news-room/fact-sheets/detail/sanitation, retrieved 22 April 2021.

The World Bank. Undated. "Citywide inclusive sanitation (CWIS) initiative." Undated. https://www.worldbank.org/en/topic/sanitation/brief/citywide-inclusive-sanitation, retrieved 25 April 2021.

World Bank / WHO / ILO/ Water Aid. 2019. *Health, Safety and Dignity of Sanitation Workers. An Initial Assessment*. Washington, DC: World Bank. https://www.who.int/news-room/detail/14-11-2019-new-report-exposes-horror-of-working-conditions-for-millions-of-sanitation-workers-in-the-developing-world.

Worster, Donald. 2017. "The good muck: toward an excremental history of China." *RCC Perspectives. Transformations in Environment and Society*, 2017 (5). https://doi.org/10.5282/rcc/8135

Xue, Yong. 2005. "'Treasure Nightsoil as if it were gold'. Economic and ecological links between urban and rural areas in late imperial Jinanan." *Late Imperial China* 26 (1): 41–71.

Zhou, Xialqin et al. 2018. "Review of global sanitation development." *Environment International* 120: 246–261.

7 Power and responsibility in transition to sustainable food packaging

Kirsi Sonck-Rautio, Sanne Bor, Greg O'Shea, Nina Tynkkynen, Vilja Varho, and Maria Åkerman

Introduction

Plastics have played an instrumental role in the development of modern society. Due to their superior material properties and affordability, polymers are widely used in different economic sectors from packaging and construction to electronics and agriculture. At the same time, the negative environmental impacts of plastics cause increasing global concern. From the viewpoint of the circular economy, the challenge is twofold: how to improve the circularity of these materials in the economy to increase resource efficiency, and/or how to replace plastics with materials whose circularity is higher than that of plastics. To meet both challenges, technological solutions are, obviously, of crucial importance, as are the actions of policymakers, businesses, and the packaging industry. Yet, it also requires action and adaptation of individual consumers and consumer groups before new solutions can become mainstream tools to promote a circular economy. It is important, however, to note that these actions and solutions are not always available for all the consumers or communities and that the power to act is not always evenly distributed.

In 2010, 5–13 million tonnes of plastic waste were estimated to have entered the oceans (Jambeck et al. 2015). One of the main reasons for the growing amount of plastic waste is disposable food packaging, the demand for which is likely to increase. In 2018, over 340 million metric tons of plastic waste was generated globally. Packaging accounted for 46% of plastic waste generation (Statista 2020). In addition to plastic, cardboard and paper, metal, and glass are also used. Waste collection and recycling systems are often inadequate, as is responsible human behaviour. At the same time, packaging is necessary for protecting food during transportation, retail, and storage. As food waste often has a greater burden on the environment than food packaging has (Silvenius et al. 2013), getting rid of all packaging or even plastic packaging is not a solution.

A food package is usually used only once. After this, it is either recycled, incinerated, landfilled, or lost in the environment. Some initiatives for reusable package solutions have emerged, but policy and industry initiatives have mainly been aimed at improving the recyclability of packages. Some materials, such as glass and metals, are technically easy to recycle. Plastic packages, however, often

DOI: 10.4324/9781003255246-7

consist of different layers or materials that are difficult to recycle. Food eaten on the go is packaged in small packages, which are particularly easily discarded into nature. Recently, the European Union has banned some plastic applications, such as drinking straws, and has set increasingly strict recycling targets. New policies and strategies recently introduced to address the problem in Europe include the EU Plastic Strategy (2018), the Single-Use Plastic Directive (2021), the EU Circular Economy Action Plan (2020), the European Strategy for Plastics in a Circular Economy (2018), and at the private-sector Ellen MacArthur initiatives. Furthermore, researchers and the packaging material value chain actors develop new materials and package types; the food industry and retail sector search for and take to use solutions that, e.g., reduce plastic use or increase the recyclability of packages; consumers pay increased attention to packaging and recycling and media and non-governmental organisations (NGOs) increase the public awareness.

One essential factor in increasing the sustainability of packaging is consumer demand and individual consumer behaviour, which, in turn, is dependent on the ways societies are built and formed. Developing sustainable food packaging must be done in a balance between ecological and economic aspects, and at the same time acknowledging consumer demands for price, performance, practicality, and perceptions of environmental impact (Nordin and Selke 2010). Consumer demand, on the other hand, is also connected to inclusivity in terms of affordability and access to different solutions. These demands are not always in line with those from the scientific Life Cycle Assessments (Steenis et al. 2017; Dilkes-Hoffman et al. 2019). The consumer role in the sustainable consumption of food packaging is particularly relevant because the circular economy of food packaging is connected to recycling, where the consumer is a key factor, given that the infrastructure for recycling is available. Yet, in the context of sustainable transition, the consumers' perspective is often ignored.

The Oslo symposium of 1994 defined sustainable consumption as the

> use of services and related products, which respond to basic needs and bring a better quality of life while minimising the use of natural resources and toxic materials as well as the emissions of waste and pollutants over the life cycle of the service or products so as not to jeopardise the needs of future generations.
>
> (IISD 1995)

In the context of food packaging, sustainable consumption thus covers products which minimise the use of energy and fossil fuels and other raw materials (e.g., water), as well as reducing waste and emission. The circular economy of food packaging is not only about sustainable consumption, but also more generally about societal and cultural values which guide environmental behaviour, such as proper disposal and recycling habits. However, the other aspects of societal, cultural, and economic factors have often been ignored. Previous research has mainly seen the role of human agents in social change and transition as attempts by individuals to change their consumption behaviour and, consequently, their lifestyle choices by following the instructions that NGOs and governments are providing (Spaargaren and Oosterveer 2010). This approach, according to social scientist Gert Spaargaren, has proven to have many shortcomings, especially with the rapidly increasing globalisation of consumption (Spaargaren 2003). The entire notion that sustainable consumption is mainly a question of individual environmental responsibility has faced criticism and is blamed for scapegoating, and the responsibilisation of consumers, consequently downplaying the role of other actors (Eden 1993; Paço and Rodrigues 2016; Akenji 2014).

In this chapter, we focus on the views that different actors and stakeholders in Finland in the packaging field have about power and responsibility regarding a transition towards more sustainable food packaging. This transition entails packaging solutions which support circular economy, supporting infrastructure and consumer values and behaviour that are in line with the principles of circular economy, including solidarity and inclusivity aspects highlighted in this volume. Finland as a case offers insights into a society that already has relatively efficient infrastructure for recycling, high living quality, and equality in society in terms of income, education, and gender. Finland can thus provide an example where the preconditions for inclusive circular economy are well met, but demonstrates how even with such prerequisites there are several pitfalls that hinder the transition to circular economy. Some of these pitfalls, as we have identified, are connected to the questions of power and responsibility. We examine how the actors perceive their own roles and responsibilities vis-à-vis others' roles and responsibilities. With the framework of power and responsibility in mind, we interviewed the actors whose roles in food packaging were recognised to be of significance during our operation environment analysis, in order to map their views on relative power and responsibility.

Responsibility and power in sustainability transition

The concept of environmental responsibility suggests that all people are obliged to enact solutions to emerging and existing environmental issues (Hobson 2006). The grounds for such an obligation can be thought of in terms of solidarity: responsibility is grounded in solidarity, which generates actions that can be described as the taking of responsibility for others, humans or non-humans; ultimately, the planet. One is capable of taking on the responsibility because the one in need of help is "one of us" (Principe 2000).

Individual environmental responsibility, on the other hand, here refers to the notion that an individual consumer has agency, capacity, and causal responsibility for addressing environmental problems (Eden 1993). This notion has long been utilised both by business and societal institutions through promoting, for example, green consumerism and recycling. Placing the burden of sustainable consumption – a term that has also been widely criticised as being an oxymoron (Bălan 2020) – on to the shoulders of the consumer raises the question of whether people fulfil their perceived individual environmental responsibility through consuming, instead of practising their political agency by voting or engaging in environmental activism (Eden 1993; Lorenzen 2014). In addition to this set of criticisms, social scientists of consumption have often interpreted individual environmental responsibility as consumer sovereignty. Consumer sovereignty assumes that consumers hold the ultimate power – and with that, also the responsibility – to make the system more sustainable (Eden 1993). Scholars have criticised businesses and even governments for advocating the notion of consumer sovereignty as a means for scapegoating consumers and creating the general notion that environmental responsibility is a consumer issue, rather than something businesses should take responsibility for (Eden 1993; Akenji 2014; Evans, Welch, and Swaffield 2017; Pekkanen 2020).

Emphasis on sustainable consumption consequently directs the focus away from recognising and addressing the roles of institutions, infrastructures, and cultures as the source of routines, habits, and beliefs. In reality, these factors guide the processes of consumption and at the same time, have power in advocating more sustainable development (Akenji 2014; Evans, Welch, and Swaffield 2017; Pekkanen 2020). Scapegoating individual consumers while downplaying the role of other stakeholders, can also diminish the sense of solidarity among the consumers. Critical voices also point out that the ability or capability of individuals to affect global markets is largely overestimated, especially when it comes to those whose socio-economic status is low (Massey 2004) – many are excluded from having a chance to "vote" with their consumption choices merely due to their economic status. However, even with this criticism in mind, it is important to be cautious to claim that consumers have no power – and thus no responsibility – as that is not the case. The role of consumers should not be excluded, but instead of focusing on the central role of human agency alone, equal focus should be targeted towards social structures (Spaargaren 2003).

In this chapter, we analyse the dimensions of power in the transition to more sustainable food packaging systems with the framework of four dimensions of power. We see these dimensions to be power *over*, power *to*, power *with*, and power *within*. The first dimension of power, *power over*, consists of A's domination over B, by making B do something which B would not otherwise have done, as it was defined by Robert A. Dahl (1957). Power over is usually interpreted normatively as domination. We use this notion of power over, however, in a little less dominative way – there is no imperative that B would act unwillingly, it is merely a question of acting in response to another actor's action (Lukes 2005). The *power to* has a dimension of agency within it – a person has the capacity to act upon something. *Power with*, on the other hand, refers to power that arises from synergies between partners or from collaboration, from the community and solidarity among stakeholders, whereas *power within* holds the notion of empowerment and inclusivity – as John Gaventa (2006) formulated it, it is not only about having agency, but about self-identity, awareness, and confidence to act (for further development of discussions regarding the dimensions of power see, e.g., VeneKlasen and Miller [2002] and Haugaard [2012]).

With power comes individual responsibility. It has been shown that individual environmental responsibility is strongest when, similarly, the individual feels that pro-environmental behaviour is effective, thus creating a sense of agency. This pre-requires feelings of inclusivity and solidarity, either towards other people or towards the planet, or both. The sense of moral obligations of responsible behaviour grows even larger if the individual perceives this agency to be more essential than those of others (Eden 1993). Similarly, Kollmus and Agyeman (2002) point out that there are several other theories regarding the importance of a sense of responsibility in environmental behaviour; in 1987 Hines, Hungerford, and Tomera argued that people's environmental behaviour is affected, among other things, by their locus of control. Locus of control is a term referring to an individual's perception of their agency, ability, and capability to advocate change through their own behavioural choices. With the same tone, James Blake (1999) argued that there is a set of three barriers, individuality, responsibility, and practicality, that constitute the gap between environmental concern and action. Blakes' second barrier stated that people sense when not having an influence on a situation, and hence feel they should not have responsibility for it either. Lack of trust in institutions is also a factor in the lack of sense of responsibility, as it also undermines the efforts of collective solidarity – or power with.

Methodology

A point of departure for our empirical data gathering was a set of workshops organised in the framework of our research project on sustainable food packaging in 2020 in Finland. During these workshops (see Table 7.1 and the following section for further information on the workshop participants, etc.), we were

Table 7.1 Data and the stakeholders presented in this study

Data		Number of participants
Power triangle – workshops	Two workshops: 1) for NGOs and 2) for businesses	7 and 30
Ministry interviews	Ministry of the Environment, Ministry of Economic Affairs and Employment, Ministry of Forestry and Agriculture	6
Businesses	Food industry, packaging industry, and material industry	20
NGOs	NGOs whose interests involve environmental issues and consumers – citizens' involvement	3
Consumers	Consumer panel which gathers online and are asked to voice their views and opinions on certain topics	22
Associations		15

mapping the stakeholders' perceptions on the power structures among the different actors in the transition towards more sustainable food packaging. During these two workshops, 37 participants participated in the power triangle exercise. Power triangles (Kohl et al. 2018) is a simple tool for mapping views about stakeholders. It uses a triangle, where a tip is facing up (Figure 7.1). Workshop participants place different stakeholders on the triangle, so that the stakeholders who have more power to affect a topic are higher (near the tip) and stakeholders with less power are nearer the bottom. It is also possible to place stakeholders outside the triangle if they are not considered to have a role in changing the issue at hand. We utilised these power triangles formulated by the participants as data in this study, but in addition to figures alone, we took notes of the discussions between stakeholders, which were also analysed for this chapter.

Adding to the power triangle workshops, we conducted 44 stakeholder interviews within different stakeholder groups. Table 7.1 presents more detailed information on the data. Among the interviewees were representatives from the packaging industry, food industry, different ministries (Ministry of the Environment, Ministry of Forestry and Agriculture, and Ministry of Economic Affairs and Employment), and non-governmental organisations. We analysed the interviews with the framework focusing on aspects of responsibility and power.

In addition to stakeholder interviews, we asked 22 consumers how, in their view, they can influence the transition towards more sustainable food packaging (including efficient circulation of materials) and what kind of responsibilities they see each actor having. The consumer panel was gathered via open invitation on our project website, distributed via email lists and social media. The panel holds 44 active participants of which 22 answered our questions regarding responsibility and power to influence.

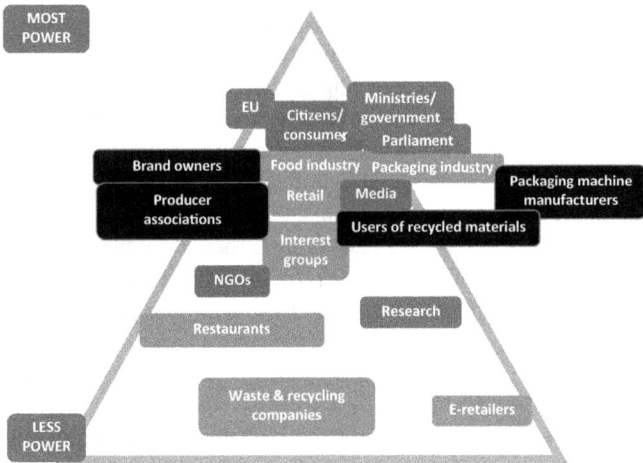

Figure 7.1 An example of a power triangle created by business representatives in a workshop.

Stakeholder views on power and responsibility

Power triangles

The first workshop was for six non-governmental organisations that focus on environmental and consumer issues. The second one was for business representatives from different sectors (food industry, packaging industry, and retail). Although the results are indicative rather than conclusive, they do paint an interesting picture of the power to influence the food packaging system.

First, the images are all rather different from one another. Second, the respondents seem to consider someone else than the group they themselves represent as being dominant. For example, the NGO representatives placed the NGOs close to the bottom of the triangle. These results indicate that many different actors do have some influence, and none obviously dominate.

Most respondents placed the European Union higher than the national parliament or government, one group considered them to have equal power. In all cases, the governance was seen to have a lot of influence. Most groups place media in the middle of the triangle but one group placed it to the top, in the most influential position. They reasoned that the media determines which topics are discussed publicly and influences policymaking and consumer choices.

The food industry and packaging industry were considered to have a lot of influence in all workshop groups. Perhaps surprisingly, waste management and recycling companies were mostly seen to have a fairly small role. The production associations, i.e., the associations that manage the collection of recyclable material on the behalf of the producers, were specifically mentioned by one group and given a higher role in the transition.

The role of consumers differs in the power triangles. Some respondents (particularly the industry representatives) give consumers more power than the industry, others see them as being much less powerful. One workshop group discussed the role of consumers and noted that while a single consumer has very little power, as a group the consumers and their perceived requirements influence the decisions made in the industry. This dualism between a single actor within a group and the group as a whole may explain more widely the differences in views. Each actor, such as a company, may see their own position as dependent on others, but as a group, each segment, such as a business sector, is influential.

In order to get a more detailed account of each stakeholders' views on power and responsibility to create a circular food packaging system, we analysed our interviews applying the responsibility and power framework presented above. Below, each stakeholder groups' interviews are described from this perspective.

Policymakers

The ministry representatives define their responsibilities slightly differently depending on the ministries' core focus areas and their role within the national governance. Following from that there were differences in what was emphasised when defining the meaning of sustainable packaging. Still, all the interviewed ministry informants recognised the role of food packages in protecting food in the food chain, to secure its safety and to avoid food loss, but at the same acknowledged the need to decrease the environmental burden of packaging. In general, the ministry informants emphasised the role of ministries in preparing regulatory frameworks, implementing legislation, setting up policy schemes and innovation funding and facilitating the collaboration between business and RDI actors.

Responsibilities that ministry respondents related to these identified tasks include, particularly, responsibility for the accuracy of the evidence used as a basis of decision making, responsibility for the health of people (food safety, nutrition), and responsibility for bringing the other actors around the same table (facilitation, orchestration). In general, it can be concluded that the ministry respondents felt responsible for taking care of the evidence base of policymaking while influencing the operational environment of food chain actors and consumers to genuinely advance sustainability and not to cause unanticipated negative outcomes either in terms of environmental or socio-economic impacts. This was regarded as a tricky task due to knowledge uncertainties related to the sustainability effects of food packaging. What was common with ministry interviews was the claim that basically, ministries cannot solve the sustainability problems themselves as the innovative solutions and their competitiveness are dependent on the acts taken by business and research actors. There were however some differences between the views of representatives of different ministries which reflect the division of labour within the government system.

The Ministry of the Environment in Finland is responsible for the national implementation of many of the EU-level governance measures addressing the reduction of plastic food packages and the increase of their recycling both as materials and reusable packages. Following from this position, and due to difficulties in measuring the recycling targets of different materials, the focus of interviews was very much on the contested facts and responsibility of the ministry to utilise science-based evidence as a basis of policies and their implementation. In addition to underlining the need to take care of the environmental impacts of food packaging and regulation, the ministry informants also expressed responsibility for the interests of national industries in the face of EU level regulation. In addition to a strong regulatory role, the Ministry of the Environment has taken the responsibility of facilitating the collaborative governance processes and development of shared goals between different societal actors (businesses, industry associations, research, development, and innovation [RDI]) to reduce the production of plastic waste and to increase the recycling of plastics as part of creating a national plastic roadmap. Following this, the Ministry of the Environment has taken the role of an orchestrator in the sustainability transition which was very much appreciated by several of our interviews with industry associations and business actors.

All the ministries emphasised that changes in food packaging require innovation and, in this sense, it is at the hands of RDI and business actors. The responsibility of ministries in promoting innovations was however a topic, which put forward contested views and showed clearly how much the existing division of work and boundaries between different ministries define their roles. In terms of innovations, the responsibility was very much put on the Ministry of Economy and Employment, which has, in general, the role of promoting innovations, employment, and growth of the export industry. Nevertheless, despite this general division of tasks between ministries, all the interviewed ministries deliver RDI funding in their own fields. Also, when talking about innovations, the informants in all the ministries addressed responsibility for the food industry and retail actors. The Ministries of the Environment, Agriculture and Forestry, and Economy and Employment have all signed from the part of the national government the voluntary material efficiency deal initiated by the food industry and retail associations and packaging industry associations. Our informants emphasised that this kind of promotion of ambitious goal setting for the private actors is one of the ways in which ministries can take responsibility for pushing businesses forward in sustainability transition.

However, when directly asked about how ministries can support the formation of new market opportunities for novel food packaging innovations which currently suffer from being not competitive in relation to existing packaging solutions, the Ministry of the Environment informants answered that market formation is not their responsibility. Their responsibility is to create new regulatory market conditions and policy goals and to communicate them clearly and after that, it is the businesses which need to take action. The representative of the Ministry of Economy and Employment, however, recognised the role of the public sector in the market creation and claimed that public procurement should be more used to share the companies' risks of introducing new products and services for the markets. In this case, the initiative was seen to be at the hands of local and regional public organisations whereas the ministries should provide guidelines for how to impact public procurement.

Businesses

The move towards sustainable food packaging involves four different industries: the materials industry, the packaging industry, the food industry, and the recycling industry. In the material industry, it is particularly the fibre industry that is seen as a potential for sustainable innovations that could replace plastics, but the chemical industry is also working to develop alternatives in terms of bio-based, more easily recyclable, and biodegradable options. The metal and glass types of packaging provide fewer opportunities, as they rely on non-renewable raw materials, and create solutions that are heavier and thus cause more transportation costs and emissions, though they are relatively well recyclable.

The packaging industry is involved in innovating ways to use and form the materials into packages which use less material, create less waste, create the barriers needed to protect the food (and prevent food loss), and create fewer emissions in logistics and storage, while still being attractive. The innovations that this industry can make to impact the environment are multiple. As actors between the materials and the food industries, they try to make what customers want with the materials available. This industry could come up with sustainable packaging solutions, help in making sure that these become more available on the market, and push material producers to innovate. Our interviews, however, indicate that they try to avoid this responsibility, and can do so, as the food industry is made responsible for the packages they put on the market. The packaging industry indicates a willingness to test and produce sustainable solutions when requested, but is against the higher prices this brings.

The food industry also plays an important role. They are making the packaging choices and considering the different pay-offs they can accept. The food industry also makes efforts to influence the material and packaging industries. We have identified five different ways in which the food industry is trying to take responsibility and influence the material and packaging industry to make this move (Bor et al. 2022). First, food companies actively signal their aims to change. They do so by formulating targets and goals in their CSR reports. From the interviews, it is clear that these signals are being picked up by the material and packaging industries, and are seen as challenging them to make changes. Second, food companies actively make changes in their packaging designs and make material choices. Over the last few years, many food companies have reduced their plastics consumption by millions of tonnes. This change in buying pattern directly affects the material and packaging industry as this means there is much more opportunity for more sustainable material and packaging solutions in the market. Third, the food companies also act as a testbed for new innovations from the material and packaging industries. They are willing to support the development of new solutions and thereby may speed up and help in further developing the innovative solutions. Fourth, food companies actively invite both material and packaging companies to participate in the redesign of their packages in their efforts to meet their sustainability goals. They thereby invite the actors from the other industries to put in an effort to find and create solutions and work towards more sustainable options. And finally, for some, particularly large international food companies, the material and packaging industries are not moving fast enough and they were frustrated with the dependence on these other industries. To overcome this dependence, they decided to move into the material and packaging industries by developing their own innovation centres for food packaging (Bor et al. 2022).

Besides the influence on the material and packaging industries, the interviews we undertook also highlight that the food industry is seeing an important role for itself in being able to influence consumers. In particular, the guidance on how to recycle packages has increased. Company informants described their realisation of their responsibility for providing the necessary information to consumers. One of the companies for example discussed the closing mechanisms for bread and sweets bags, the metal structure with plastic coating. As it looks like plastic, most people tend to put it to the plastics recycling, however, it is best recycled with metal. The company realised that the consumer does not know what the package is made of and where it best is recycled and thus food companies have a responsibility to communicate this to the consumers.

Finally, the recycling industry is playing an important role via the development of sorting and recycling technologies, but they also play a key role in providing information to packaging and food companies about the current and future recyclability of materials.

Thus far, the transition towards sustainable food packaging has not resulted in much solidarity among the business actors, this is, however, understandable as transitions cause changes which provide benefits to some, while others are losing. While different actors do work together on projects, many of the companies interviewed highlighted their fear of breaking competition laws when working with competitors to find solutions, creating a further barrier to solidarity and collective action.

Non-governmental organisations (NGOs)

NGOs recognised their environmental power as being mainly about educating consumers about good environmental choices based on objective scientific facts and/or regulations and directions provided by governmental institutions. The NGO representatives' interviews did, however, have different views on ideological education, as some NGOs were more ideologically based than others. Related to educating people, influencing schoolchildren in the form of school visits, etc. was seen as a rather efficient way to promote sustainable behaviour. NGOs did note that their audience is most often people who are already somewhat aware and educated, or at least open to learning, so NGOs' lack of power lies in the fact that their ability to reach and influence all people is very limited.

Although individual consumers might have little power to influence other stakeholders, mainly producers and policymakers, NGOs considered their responsibility to transmit the voice of consumers as a group. Hence, this responsibility is based on a strong sense of solidarity. The NGOs can communicate to policymakers that people are ready and willing to adopt new policies, and on the other hand, let businesses know the wishes and preferences of "green" consumers and encourage them to take actions towards more sustainable solutions.

In terms of sorting and recycling, the consumers' role was seen as very significant, but at the same time, it was acknowledged that sorting and recycling should be made as easy, simple, and inclusive as possible. Consumers cannot act sustainably if there is no sustainable choice or if they do not know how to act. The interviewees emphasised that, for example, instructions for sorting waste should be simpler by further standardising labelling in both packaging and disposal containers. This type of infrastructural responsibility was placed on policymakers (legislation and regulations) or on business stakeholders (volunteer agreements such as green deals).

Consumers

We asked 22 consumers in what way they consider themselves responsible for consuming sustainably and for advocating effective circulation of food packaging material.

Consumers considered their power in the transition towards more sustainable food packaging to lie within their capability to impact as a group. However, as individuals, their power is lacking in terms of having an impact on the packaging industry or even policymakers. Consumers perceive that it is not in their power to influence the packaging choices the producers make. Even though consumers can "vote with their choices", they can only choose from what is offered, and therefore it is producers' responsibility to make sure their choices are the best possible ones in terms of sustainability. Many of the consumers thought their channel for influence was consumer feedback, but doubted its efficiency.

All our respondents felt it is in their power to sort and recycle. Consumers felt, rather unanimously, that it is the consumers' responsibility to sort and recycle their own food packaging waste. However, they thought that it is not their responsibility, nor is it even in their power, to organise an efficient circulation of the material, nor are they responsible for maintaining or cleaning the recycling stations. They see that creating the infrastructure should be up to the manufacturers, food industry, or the municipality, who are also responsible for making it as easy as possible for the consumer to recycle efficiently – i.e., recycling bins should be big enough, there should be more of them and closer by, and even the people with no cars should have easy access to them (not only in the supermarket parking lots).

Notably, consumers thought that they do have the possibility to promote a sustainable packaging industry by purchasing packaging that is easily recycled, but they did not mention this behaviour as their responsibility. This sense of responsibility is also lessened by the notion that for the consumer, assessing the sustainability of food packaging is very difficult, as there are so many mixed signals of sustainability. However, it should be the food and packaging industries' responsibility to manufacture packaging that is easily recycled and made of material that is environmentally friendly, as consumers cannot make decisions over which type of packaging the food will be packed.

Discussion

Based on the analysis above, the power of different actors using a framework with four types of power – power over, power to, power with, and power within – can be discussed as follows.

Power over was defined as the ability of one actor to prevail over another actor. Our results indicate that there is no obviously dominating actor within the food packaging system, nor are there any actors with no power at all. Ministries seem to have power over other actors in terms of regulations and legislation, creating the preconditions for sustainable food packaging industry and consumption. However, according to our material, the food industry has power over other actors in creating and promoting a sustainable market, though limited, as they remain dependent on the material and packaging industries for the sustainable innovations needed. Consumers as a collective were seen as having power over all actors. On one hand, consumer demand guides the market – what they are buying is guiding to a degree what is put on the market. It is also important to note that being able to vote with one's choice is often not an option for those with low income. On the other hand, consumers are, through active participation in societal discussion, affecting the policies and regulations being made.

All actors have the *power to* act, although each has different strengths. Ministries have the power to fund innovation, for example, whereas material and packaging industries have the power to create innovations. NGOs have the power to educate consumers and have the power to influence policymakers and, in some cases, the industry as well. Consumers have the power to recycle materials, albeit they do not have the power to influence the recycling infrastructure. Being able to recycle can thus also be seen as an issue of equal opportunities. Not everyone has equal access to recycling bins due to their location or possibilities and facilities to sort and store the waste in their homes. There is, for example, a great difference whether one lives in a house or in an apartment building, in a city centre or remote countryside. Consumers noted that even though as a group they have the power to do many things, they are not well organised and therefore do not have the possibility to control or steer their message. Also, consumers do not form a single homogeneous group, but are very heterogeneous in regards to their preferences, ideologies, needs, etc. As individuals they at large lack *power to.*

Power with follows from synergies between partners or from collaboration. There are many examples of potentially fruitful synergies in terms of sustainable food packaging. For example, ministries funding innovations and material and packaging industries using these funds to innovate; consumers collaborating with NGOs to make their voice heard and voluntary agreements within the industry, to name but a few. These synergies create a potential for solidarity. The fourth dimension of power, *power within*, is connected to empowerment and agency, awareness and capability, which is, as established previously, closely connected to (individual) environmental responsibility and the underlying sense of solidarity and inclusivity. Those stakeholders who are aware of ways they can act, have the capability of acting, and feel their agency is equal to or even more important than that of others, indicating a strong solidarity aspect, also have a greater sense of environmental responsibility.

The actors examined in this chapter had different takes on responsibility. Ministries' responsibility, as defined by the representatives of different ministries, was to facilitate transformation and produce accurate knowledge of sustainability. Food industry respondents also brought up the question of them having a responsibility to educate and inform the consumers of their products, namely the proper ways of sorting and recycling them. NGOs' responsibility was seen predominantly as educating consumers, but also as communicating consumers' preferences to other actors, i.e., acting as a mediator of some sort. The third responsibility the NGOs were seen as having was securing the interests of the consumers in the transition processes, by demanding better recycling instructions and actions against greenwashing, for example. Consumers had a tendency to think that individually they have no power over other actors, nor any significant power to act upon to transition towards more sustainable food packaging, and so they perceive the lack of environmental responsibility. A clear exception was consumers' attitude towards recycling, which they felt they have both power and responsibility to do.

However, from the consumers' point of view, it was other actors' responsibility (policymakers, producers) to build an accessible and efficient infrastructure so that consumers have no excuse not to recycle. Interestingly, among all actors, it was perceived that consumers have power over other actors, as well as the power to act, yet according to our material, they have no responsibilities as a group. This could be due to the lack of a forum or channel which would, on the other hand, execute the power they have, and with that, prompt the sense of responsibility.

Conclusions

In discussions with different food packaging stakeholders, it has become clear that one of the main challenges in behaving responsibly is the uncertainty of what is sustainable. For the industry, the lack of information on overall sustainability – especially since the sustainability of certain packaging solutions is highly dependent on the product the packaging is supposed to cover – is clearly a barrier for product development and innovation. For policymakers the lack of clear sustainability assessments is a barrier to making the best possible policies and, finally, for the consumers, the lack of knowledge clearly constrains their sustainable behaviour as is it very hard for them to assess the sustainability of each packaging. The lack of knowledge, information, and education can decrease the sense of responsibility as well as the underlying solidarity in all stakeholders, as it has an impact on the attributes – such as awareness and capability – having power within. In our data, researchers were often mentioned as a group that would need to produce such information. It can be said that the research community has an important responsibility in providing science-based evaluations of different materials and solutions. The difficulty often lies in the life-cycle of packages: the sustainability depends much on the recycling and reuse possibilities of each location.

From our material, we find that in terms of power and responsibility in the transition towards more sustainable goods packaging solutions, there is no stakeholder that has clearly more power or more responsibility than the others. However, there were indicators to suggest that different stakeholder groups are dependent on each other – i.e., the industry is dependent on the regulatory frames policymakers create, and on the funding they provide. At the same time, policymakers cannot create markets for more sustainable products, namely those supporting a circular economy. Consumers, albeit they have the power to recycle, are dependent on the recycling infrastructure provided. Similarly, although collectively consumers can "vote" with their consumption choices, individual consumers are dependent on the choices the industry and policymakers make available to them, in terms of packaging selections in stores, for example. The sense of responsibility or solidarity does not directly help if the infrastructure is lacking and/or there is no possibility to choose. Indirectly, the sense of responsibility and solidarity can, naturally, help to put pressure on industry and policymakers, particularly if consumers are acting as a group.

Since there is a clear distinction between the sense of power and responsibility of the consumer, whether consumers are seen as individual agents or as a group, there is an evident niche for an institutional organisation which would represent consumers as a group but would also act *with* consumers, thus creating power *within* and consequently, increasing the sense of environmental responsibility and solidarity. For power and responsibility to be distributed equally and efficiently in the food packaging sector, there is an imminent need for evidence-based sustainability recommendations for all stakeholders and more efficient cooperation, where an assigned representative body would be in charge of representing consumers – and also in developing ways to channel the concerns and initiatives of individual consumers.

Acknowledgement

The authors wish to acknowledge financial support from the Package-Heroes project (grant numbers 320215, 320217, 320298, and 320299) funded by the Strategic Research Council of the Academy of Finland.

References

Akenji, Lewis. 2014. "Consumer Scapegoatism and Limits to Green Consumerism". *Journal of Cleaner Production* 63 (January): 13–23. https://doi.org/10.1016/j.jclepro.2013.05.022.

Bălan, Carmen. 2020. "How Does Retail Engage Consumers in Sustainable Consumption? A Systematic Literature Review". *Sustainability* 13, no. 1 (December): 1–25. https://doi.org/10.3390/su13010096.

Blake, James. 1999. "Overcoming the 'Value-Action Gap' in Environmental Policy: Tensions between National Policy and Local Experience". *Local Environment* 4, no. 3 (May): 257–78. https://doi.org/10.1080/13549839908725599.

Bor, Sanne, Greg O'Shea, Henna Sundqvist-Andberg, and Maria Åkerman. 2021. Moving to Sustainable Food Packaging: The actions of the food industry. In: Proceedings of the 35th Annual Conference of the British Academy of Management, Lancaster: British Academy of Management, 539.

Dahl, Robert A. 1957 "The Concept of Power". *Behavioral Science* 2, no. 3 (January): 201–215. https://doi.org/10.1002/bs.3830020303.

Dilkes-Hoffman, Leela, Peta Ashworth, Bronwyn Laycock, Steven Pratt, and Paul Lant. 2019. "Public Attitudes towards Bioplastics: Knowledge, Perception and End-of-life Management". *Resources, Conservation and Recycling* 151 (December): 1–8 https://doi.org/10.1016/j.resconrec.2019.104479.

Eden, Sally E. 1993. "Individual Environmental Responsibility and Is Role in Public Environmentalism". *Environment and Planning A: Economy and Space* 25, no. 12 (December): 1743–1758. https://doi.org/10.1068/a251743.

Evans, David, Daniel Welch, and Joanne Swaffield. 2017. "Constructing and Mobilizing 'the Consumer': Responsibility, Consumption and the Politics of Sustainability". *Environment and Planning A: Economy and Space* 49, no. 6 (February): 1396–1412. https://doi.org/10.1177/0308518X17694030.

Gaventa, John. 2006. "Finding the Spaces for Change: A Power Analysis". *IDS Bulletin* 37, no. 6 (November): 23–33. https://doi.org/10.1111/j.1759-5436.2006.tb00320.x.

Haugaard, Mark. 2012. "Rethinking the Four Dimensions of Power: Domination and Empowerment". *Journal of Political Power* 5, no. 1 (April): 33–54. https://doi.org/10 .1080/2158379X.2012.660810.

Hines, Jody M., Harold R. Hungerford, and Audrey N. Tomera. 1987. "Analysis and Synthesis of Research on Responsible Environmental Behavior: A Meta-Analysis". *The Journal of Environmental Education* 18, no. 2 (July): 1–8. https://doi.org/10.1080 /00958964.1987.9943482.

Hobson, Kersty. 2006. "Environmental Responsibility and the Possibilities of Pragmatist-orientated Research". *Social & Cultural Geography* 7, no. 2 (April): 283–298. https:// doi.org/10.1080/14649360600600734

International Institute for Sustainable Development (IISD). 1995. *Oslo Roundtable on Sustainable Production and Consumption*. New York: The International Institute for Sustainable Development (IISD).

Jambeck, Jenna R., Roland Geyer, Chris Wilcox, Theodore R. Siegler, Miriam Perryman, Aanthony Andrady, Ramani Narayan, and Kara Lavender Law. 2015. "Plastic Waste Inputs from Land into the Ocean". *Science* 347, no. 6223 (February): 768–771. https:// doi/10.1126/science.1260352.

Kohl, J., N. Wessberg, M. Dufva and S. Kivisaari. 2018. "Tools and Approaches Creating Shared Understanding of Systemic Change: Reflections on Three Case Studies". *VTT Technical Research Centre of Finland Ltd* 323, no. (February): 1–45. https://www .vttresearch.com/sites/default/files/pdf/technology/2018/T323.pdf

Kollmuss, Anja and Julian Agyeman. 2002. "Mind the Gap: Why Do People Act Environmentally and What Are the Barriers to Pro-environmental Behavior?" *Environmental Education Research* 8, no. 3 (July): 239–60. https://doi.org/10.1080 /13504620220145401.

Lorenzen, Janet A. 2014. "Green Consumption and Social Change: Debates over Responsibility, Private Action, and Access: Green Inequality". *Sociology Compass* 8, no. 8 (August): 1063–1081. https://doi.org/10.1111/soc4.12198.

Lukes, Steven. 2005. *Power: A Radical View* (Second ed.). Basingstoke: Palgrave Macmillan.

Massey, Doreen. 2004. "Geographies of Responsibility". *Geografiska Annaler: Series B, Human Geography* 86, no. 1 (November): 5–18. https://doi.org/10.1111/j.0435-3684 .2004.00150.x.

Nordin, Norbisimi, and Susan Selke. 2010. "Social Aspect of Sustainable Packaging". *Packaging Technology and Science* 23, no. 6 (September): 317–26. https://doi.org/10 .1002/pts.899.

Paço, Arminda, and Ricardo Gouveia Rodrigues. 2016. "Environmental Activism and Consumers' Perceived Responsibility". *International Journal of Consumer Studies* 40, no. 4 (January): 466–474. https://doi.org/10.1111/ijcs.12272.

Pekkanen, Tiia-Lotta. 2020. "Institutions and Agency in the Sustainability of Day-to-Day Consumption Practices: An Institutional Ethnographic Study". *Journal of Business Ethics* 168, no. 2 (January): 241–260. https://doi.org/10.1007/s10551-019-04419-x.

Principe, Michael A. 2000. "Solidarity and Responsibility. Conceptual Connections". *Journal of Social Philosophy* 31, no. 2 (Summer): 139–145. https://www.academia .edu/45378792/Solidarity_and_Responsibility_Conceptual_Connections?from=cover _page

Silvenius, Frans, Kaisa Grönman, Juha-Matti Katajajuuri, Risto Soukka, Heta-Kaisa Koivupuro, and Yrjö Virtanen. 2013. "The Role of Household Food Waste in Comparing Environmental Impacts of Packaging Alternatives". *Packaging Technology and Science* 27, no. 4 (June): 277–292. https://doi.org/10.1002/pts.2032.

Spaargaren, Gert. 2003 "Sustainable Consumption: A Theoretical and Environmental Policy Perspective". *Society & Natural Resources* 16, no. 8 (January): 687–701. https://doi.org/10.1080/08941920309192.

Spaargaren, Gert, and Peter Oosterveer. 2010. "Citizen-Consumers as Agents of Change in Globalizing Modernity: The Case of Sustainable Consumption". *Sustainability* 2, no. 7 (June): 1887–1908. https://doi.org/10.3390/su2071887.

Statista 2020. "Distribution of Plastic Waste Generation Worldwide in 2018, by Sector". Accessed 26.8.2021. https://www.statista.com/statistics/1166582/global-plastic-waste-generation-by-sector/.

Steenis, Nigel D., Erica van Herpen, Ivo A. van der Lans, Tom N. Ligthart, and Hans C.M. van Trijp. 2017. "Consumer Response to Packaging Design: The Role of Packaging Materials and Graphics in Sustainability Perceptions and Product Evaluations". *Journal of Cleaner Production* 162 (September): 286–98. https://doi.org/10.1016/j.jclepro.2017.06.036.

VeneKlasen, Lisa and Valerie Miller. 2002. "Power and Empowerment". *PLA Notes* 43: 39–41. https://pubs.iied.org/sites/default/files/pdfs/migrate/G01985.pdf?

8 Throwaway culture and the circular economy

Lifespan concepts in regulation

Taina Pihlajarinne

Introduction

The sustainable lifespan of products and materials is a crucial target as part of the circular economy (CE). This term means a new approach to economics that aims to set in place a system where the value of materials is utilised in the economy for as long as possible, in contrast to the linear (take, make, dump) type of economy. CE aims to reduce waste and optimise the use of resources, with a focus on reuse, repair, recycling, functional economy, eco-design, industrial ecology, sustainable supply, and responsible consumption (Gallaud and Laperche 2016: 2–5). Broader definitions reflecting a wide scope of action are applied as well.[1] CE is seen as having great potential in leading to radically new practices and helping states achieve increased sustainability at low costs (Ghisellini, Cialani, and Ulgiati 2016: 12).

This chapter discusses the relationship between consumer behaviour patterns relating to CE activities and "expected" or "normal" lifespan assessments as part of regulation. These kinds of thresholds can be found, for instance, in the area of private law regulation. In this context, I use consumer law and intellectual property law norms as examples. These types of concepts are especially interesting since they operate at the interface between requirements set by legal norms and real-life consumer patterns and attitudes created as a part of Western consumer cultures. Additionally assessed is the question of how to promote sustainability in this specific context of formulation and application of these types of norms. Specific focus is on the Nordic countries.

A traditional CE strategy has been to improve the eco-efficiency of products, aimed at more extensive production with fewer resources and less waste creation (Boulanger 2010: 4–5). This strategy must, however, be combined with a so-called sufficiency strategy, that is to say, reduced material and product consumption.[2] Reducing consumption involves methods for lifespan extension such as repair and recycling, presumably leading to decreased consumption. Second-hand products have a similar effect.

Product lifespans are currently under many kinds of regulatory interventions. A more sustainable lifespan of products is an explicit target for regulation in many areas. The EU is currently active in the CE area. The Waste Directive (Directive 2008/98/EC of the European Parliament and of the Council of 19 November

DOI: 10.4324/9781003255246-8

2008 on waste and repealing certain directives) sets some basic waste management principles and a waste management hierarchy where prevention of waste is the first priority; it also mentions extending the lifecycles of products as one method of achieving this (see recital 12 of the directive). However, if the chosen CE approach relies too heavily on waste management, the transition towards CE fails.[3]

The Ecodesign Directive (Directive 2009/125/EC of the European Parliament and of the Council of 21 October 2009 establishing a framework for the setting of ecodesign requirements for energy-related products) framework improves the energy efficiency of products but it also includes product-specific rules, for instance on their durability or availability of spare parts and repair information. However, the scope of these rules is quite restricted, for instance, ecodesign repair-related requirements so far apply only to certain specific household appliances and professional repairers. The CE action plan (CEAP) (a new Circular Economy Action Plan for a cleaner and more competitive Europe, COM/2020/98 final) proposes new measures. According to the CEAP, the Commission will consider "establishing sustainability principles" and other bases for an extensive list of topics. The action plan includes, for example, widening the ecodesign framework to promote product durability and the right to repair, and revising consumer law to include a requirement to inform consumers of product lifespans. Action is especially focused on some product categories such as electronics, ICT, and textiles. Additionally, the plan recognises IPRs as a potential obstacle to repairs; indeed, the subsequently published Intellectual Property Strategy mentions revising design law and addressing the question of how to ensure that repair and re-use are not blocked by unfair or excessively restrictive IP practices.

Consumer behaviour patterns have a connection with the discussion about cultural aspects and sustainable development, if the definition and role of culture are understood widely. In that case, broader cultural changes towards a more sustainable lifestyle are seen as a crucial foundation for sustainable development.[4] In terms of the development of product lifespans, consumer attitudes play a central role. These attitudes are reflected in what kind of lifespan consumers expect – serving as either a potential incentive or as an impediment to producing products with a longer lifespan. They are also highly relevant in the choices that are made between repairing an existing product or buying a new one – and also in how long a consumer is willing to use a product overall. The availability of relatively cheap products due to efficient mass production as a consequence of industrialisation and economic growth, in conjunction with a common increase of living standards, has generated the so-called culture of waste. This means that, with a view to lengthening the lifespan of products, it is crucial to find mechanisms to shape consumer behaviour patterns and practices.

Despite strong involvement by the EU legislator in CE and lifespan regulation, it is important to pay attention to the overall regulative framework concerning lifespans. Therefore it is crucial also to assess norms whose aim is not directly to promote sustainability or to lengthen lifespans, but that de facto impact on expectations of acceptable lifespans in our society and in consumer attitudes. These norms might either be based on EU law or they can be non-harmonised, like some of the norms discussed in this chapter. Indeed, there is remarkable potential for creating incentives for sustainable lifespan in many areas of law where lifespan considerations for sustainability reasons are not central or that are not traditionally affected by the circular economy. So, these kinds of norms have a strong role to play for CE as well even if they are often not at the centre of the discussion on promoting sustainability. For the part of this type of norms that are harmonised by the EU, it is also crucial to recognise that Article 11 of the Treaty of the Functioning of the European Union (TFEU) states that "environmental protection requirements must be integrated into the definition and implementation of the Union's policies and activities, in particular with a view to promoting sustainable development".[5] This clause obliges the EU to take environmental objectives into account in all issues, including those areas that do not directly concern environmental protection (Wiesbrock and Sjåfjell 2014: 1). It is crucial that, since most areas of law include elements that impact sustainability, the requirement of sustainability should be involved coherently in all of them.

Consumer behaviour patterns, lifespans, moral norms, and legal norms

Some research suggests that consumer attitudes, that is to say, a culture of consumption that is characterised by lack of interest in and awareness of sustainability, serves as a main barrier for CE. Additionally, company culture that is often stuck in ongoing business models is another crucial barrier (Kirchherr et al. 2017). Research on EU Member State consumer attitudes demonstrates that on a general level, consumers are willing to engage in CE activities. However, this willingness is not yet fully actualised. The majority of consumers repair products (64%), but 36% had not done so; 90% of consumers had no experience of renting or buying second-hand products. Reasons might include consumers' lack of information on product durability and repairability as well as undeveloped markets. For instance, consumers might hesitate to undertake repairs if the use of repair services requires much effort (European Commission 2018: 10). As a possible remedy, the idea that regulatory intervention is needed is widely supported (Kirchherr et al. 2017).

In general, consumer behaviour from the sustainability standpoint is a quite widely researched area. In the literature, various approaches and perspectives have been taken. Jackson (2005) has divided them into four categories. The first group includes rational choice models such as rational choice theory, consumer preferences theory, and Lancaster's model. These are based on the assumption of consumers that make rational choices based on calculations of the benefits and costs of each choice and reach a decision on the basis of the expected net benefit. Self-interest is the factor that directs the decision. The second group is based on theories representing extensions of rational choice models, by adding perspectives that are focused on finding psychological factors (attitudes, beliefs, values) that distract the consumer from making rational choices, such as the theory of reasoned action, the theory of planned behaviour, the means-end chain theory, and the simple expectancy value-theory. The third group consists of theories offering a cultural approach to understanding consumption, the assumption that consumer behaviour is dependent on "social logic"; the individual's behaviour must be set in a social context. The focus here is on such issues as the dynamics of consumer actions, consumers' lifestyles and identities, the marketplace and cultural meanings. The fourth model uses a socio-material approach focusing more on practices than behaviour (Jackson 2005; see also Camacho-Otero, Boks, and Pettersen 2018: 4–5).

Research under these four groups demonstrates that diverse factors impact consumer behaviour in terms of lifespan expectations. Differences between countries may reflect cultural differences, history, values, markets, concrete local environmental problems, as well as government action. Differences might occur between consumers' attitudes and patterns in a particular area, region, or country, but there might also be great diversity among consumer groups irrespective of which country they belong to. Questions of consumers' lifespan expectations and attitudes in and among various countries and consumer groups are difficult to assess due to remarkable differences in product categories and research methods. Due to the variety of research methods and choices made, research conducted on these themes is not easily comparable. Additionally, diverse definitions of what is meant by lifespan might produce different results: for instance, expected lifespan can mean different things, such as the intended or ideal predicted lifespan (Gnanapragasam et al. 2017: 465).

While generally a lengthening of product lifespan is clearly desirable from an environmental perspective, this is not always the case. For instance, a very long lifespan for cars might cause an excessive amount of carbon dioxide emissions in comparison with newer eco-friendly cars. Therefore, a research result that in Finland consumers had the highest expected lifespan (22 years) for cars among the European countries involved in the research (Austria, Denmark, France, Germany, Ireland, Italy, Netherlands, Spain, Switzerland, Great Britain), 13–18 years (Oguchi and Fuse 2015: 1740) is not very positive for Finns from an environmental perspective. Additionally, it might not say very much about Finnish consumers' attitudes about lifespans in general as the result is probably caused by tax policy relating to issues reflected in car prices.

Despite these restrictions and contradictions, some general conclusions can be drawn from research on consumer attitudes towards CE in one country or several[6] or product-specific research on consumer expectations about lifespans. An interesting example is offered by European Commission research on consumer attitudes towards circular economy actions, revealing some Member State-based differences. Consumers from the former communist Eastern European countries expected a longer lifespan from products, and they were more willing to repair products. This was assumed to be caused by the historical fact that due to a shortage of commodity goods in the former communist states, product lifespan was longer than in Western societies. This is still visible in consumer patterns in those countries (European Commission 2018: 110).

Also, for instance, deforestation in a particular country correlates with higher environmental consciousness and higher recycling activity. For instance, differences in citizens' recycling activities among EU countries were correlated with citizens' activity in environmental associations. Results of CE activities on a personal level were impacted by whether or not they believed the government of the country was making reasonable efforts in terms of environmental protection (Guerin, Crete, and Mercier 2001: 195–218).

A study commissioned by the EC (2011) revealed that consumers' willingness to buy second-hand varied between the Member States: the proportion of consumers that are in general willing to buy second-hand products ranged from Slovakia, Czech Republic, and Cyprus (40%, 44%, and 46%) to Sweden, Finland, and Denmark (87%, 86%, and 83%) which were clearly above the average (68%). General willingness in the Baltic countries was on both sides of the average: Estonia 75%, Latvia and Lithuania both 67%. Product-specific outcomes demonstrated that a willingness to buy second-hand textiles was above the average (36%) in the Nordic countries (Finland 64%, Sweden 54%, and Denmark 49%) as well as in the Baltic countries (Estonia 60%, Latvia 51%, and Lithuania 46%). The countries where willingness was under the average were for instance Germany (35%), Italy (24%), and Cyprus (13%). The average willingness in Member States was generally higher (56%) concerning second-hand furniture: in the Nordic countries this willingness was again clearly above the average (Sweden 82%, Finland 78%, and Denmark 72%). However, the Baltic countries were ranked under the average (Estonia 54%, Lithuania 51%, and Latvia 42%). Willingness to buy second-hand electricity was in line with these results concerning the Nordic countries, since the rates were above the EU average (45%) in Sweden (55%), Denmark (52%), and Finland (50%). Again, the rates were under the average in the Baltic countries: Latvia (39%), Lithuania (38%), and Estonia (37%). This research indicates that attitudes are not only country-specific but also product-specific. Nordic consumers' stance here looks more positive for the circular economy in comparison with other countries. However, for some reason in the Baltic countries there seems to be a more positive attitude towards second-hand textiles compared with other product categories in this research, while generally in the EU willingness to buy second-hand textiles ranked lowest among the product categories investigated (24–26). There is a slight variation in reasons for not buying second-hand products: in Luxembourg, and Finland, the reason for not buying second-hand products was expected (lower) quality and usability, while in Ireland, Cyprus, the UK, Greece, and Hungary the reasons placed more emphasis on health and safety concerns (27–28).

However, research on Nordic CE policies commissioned by SB Insight (2019) concluded that the Nordics, except Finland, are lagging behind some other EU countries such as the Netherlands and France in terms of legislation, policies, and incentivising mechanisms.[7] The research revealed that the most popular CE actions in the Nordic countries were repair, recycling, and utilisation of second-hand products. The most important incentive for repairs for the Nordic consumer is, however, saving money, and environmental benefits serve only as a second important incentive. Accordingly, the cost of repairs was considered as a crucial barrier for repairs, as well as the fact that it is easier to buy a new item. Not surprisingly, retaining ownership of items was considered very important from Nordic consumers' perspective (SB Insight 2019: 44, 52). In all the Nordic countries consumers think that the consumer bears the main responsibility for promoting CE. However, in Norway, due to an extensive public sector, consumers believe more in the problem-solving ability of the state and therefore, in comparison with the other Nordic countries, they assign more responsibility to the public sector in CE, too (40).

The Nordics have traditionally been viewed as having high environmental awareness and technological development. This seems to be demonstrated in part by the above research results as far as consumer attitudes are concerned. However, lack of monetary incentives and supporting legislation seem to be a problem in the Nordic countries as elsewhere.

Some evidence can be found of a positive correlation between incomes and education and the average use period of products (Wieser, Troeger, and Hübner 2015). Other research indicated that socio-economic status in terms of education and income had only a modest impact on recycling activities (Guerin, Crete, and Mercier 2001: 195–218). However, a third research study indicates that the relationship between incomes and attitudes towards CE activities might not be that straightforward. It concluded that among the Nordic countries, in Norway and Denmark no correlation whatsoever existed between incomes and willingness to reduce consumption. However, both in Sweden and in Finland such a correlation existed, but increased incomes had the opposite effect in these two countries: in Sweden the lower the income, the greater the willingness to reduce consumption, while in Finland the greater the income, the more positive the attitude to reducing consumption (SB Insight 2019: 39). Therefore it seems that while the level of education might play a role in consumer behaviour relating to product lifespan, the role of incomes is far more contradictory.

The connection between consumer patterns and attitudes and moral, social, and legal norms is relevant when assessing endeavours towards lengthening product lifespans. There are diverse definitions of social and moral norms, but usually a social norm exists when a sufficient number of individuals recognise and follow it in certain situations, and expect that a sufficient number of other people will follow it in the same situations (Bicchieri 2005: 2). Moral norms are considered to be followed unconditionally (as opposed to the conditional following of social norms) and they are assumed to have an emotional effect (Miliute-Plepiene et al. 2016: 41). Research on household recycling behaviour indicates that moral and social norms are highly relevant in this context (see Miliute-Plepiene et al. 2016: 41). Due to the relative nature of the various CE acts where the aim is to keep materials in use as long as possible, moral and social norms presumably play a crucial role in the context of other CE-related activities as well. Therefore, consumer patterns and attitudes might reflect the social and moral norms on CE that, for instance, set conditions on how long a product should be kept in active use. Legal norms and social norms play a remarkable role in the process of activating moral norms. Legal norms often shape initial consumer attitudes and behaviour.

In the long run, legal norms have a tendency to be transformed into moral norms, so that internal sanctions gain more relevance compared with external sanctions (42, 48).

Legal norm thresholds for normal or expected lifespan

As explained, the focus of this chapter is on the thresholds set by legal norms regulating product lifespan by using definitions referring to the "normal" or "expected" lifespan. Here, I use examples from consumer law and intellectual property law, and it is noteworthy that neither of these areas of law is explicitly aimed at promoting environmental sustainability. However, as explained in the introduction, even though their focus is elsewhere than in environmental protection, these norms are important in shaping expectations for acceptable product lifespans in our society.

Consumer law is a traditional tool in attempting to regulate product lifespan. However, its first priority aim is to protect consumers from abuse, that is to say, their monetary interests and safety. The focus of consumer law regulation is on risk allocation between consumers and producers (see, for instance, Howells, Twigg-Flesner, and Wilhelmsson 2018: 10–11). However, consumer law has side effects that are relevant from the environmental perspective: consumer law has the potential to promote repairs and to create incentives for products with a longer lifespan. It should be noted, however, that there are also opposing arguments in that a very high level of protection for consumers buying new products might decrease their willingness to buy second-hand products (Schlacke, Tonner, and Gawel 2015: 153).

Consumer law is a widely harmonised area of law in the EU, and the Consumer Sales and Guarantees Directive (EC Directive 1999/44/EC of the European Parliament and of the Council of 25 May 1999 on certain aspects of the sale of consumer goods and associated guarantees) sets requirements for Member State legislators. Article 2(2)(d) states that goods must show the quality and performance which are normal in goods of the same type and which the consumer can reasonably expect, given the nature of the goods and taking into account any public statements on the specific characteristics of the goods made by the seller, the producer or their representative, particularly in advertising or labelling. While it is not explicitly mentioned, this might also include matters of durability as part of consumers' reasonable expectations as to the quality of the goods (Howells, Twigg-Flesner, and Wilhelmsson 2018: 183). The EU consumer law concept of consumer reflects an average consumer who is assumed to be careful, which as a standpoint has been criticised (27–29).

The directive sets a minimum period for a guarantee. It includes the requirement of the seller being liable if lack of conformity becomes apparent within two years as from delivery of the goods (Article 5.1). Member States may impose more stringent provisions including a longer period for liability, to ensure a higher level of consumer protection (Article 8.1). Additionally, under Article 7.1 Member States may allow a one-year liability period for second-hand goods. As

Maitre-Ekern and Dalhammar (2019: 411) put it, the fixed two-year guarantee period to which most Member States are committed corresponds to consumer expectations in a throwaway society. This requirement is far from being ambitious.

Due to minimum harmonisation, the Member States have adopted several kinds of solutions.[8] These can be roughly divided into three models, the first being a model with fixed liability periods, either concerning all products or diverse fixed liability periods for specific product categories. The period might be the minimum set by the directive (two years) or longer. For instance, Sweden has a three-year guarantee period for all products.[9]

A second type of model combines elements from fixed periods and the "normal" or "expected" lifespan type of assessment. For instance, in the UK and Ireland, the maximum guarantee period is six years, but the period of responsibility depends on assessment of each case from the perspective of the requirement that the quality of the product must be satisfactory. Quality is satisfactory if a product meets the standard that a reasonable person would consider satisfactory, taking account of any description of the goods, the price or other consideration for the goods, and all other relevant circumstances (UK Consumer Rights Act 2015: section 9, 1–2). In (EFTA member) Norway a two-stage system applies, with a five-year maximum for products with a longer expected lifespan.[10]

The third model is based "purely" on a more flexible "normal" or "expected" lifespan assessment. For instance, in Finland, a good is defective if it does not correspond to what a consumer ordinarily may expect in the purchase of such goods as to durability and otherwise (Finnish Consumer act sections 2, 2[5], and 4). The assessment is based on objective grounds and for instance price plays a role when assessing durability and other features as well as consumers' reasonable expectations as this is assumed to reflect quality differences.[11] Price serves as one factor that reflects quality differences – "low price, non-sustainable lifespan" therefore serves as an acceptable strategy for producers. In the Netherlands, a good is defective if it does not correspond to the consumers' reasonable objective expectations as to the features of goods in normal use (Burgerlijk Wetboek section 7 art. 17.2). Assessment of reasonable objective expectations often relies on various organisations, such as consumer and sector organisations, and their published assessment on the average or assumed lifespans for different product categories (Tonner and Malcolm 2017: 20). In Belgium, revision of the two-year guarantee period has been proposed with the aim of reducing planned obsolescence and promoting CE. The proposed guarantee would be linked with the expected lifespan for each product category, which would be defined by the state authority (SPF Économie) (Proposition de loi modifiant le Code civil et le Code de droit économique, visant à lutter contre l'obsolescence programmée et l'obsolescence prématurée et à augmenter les possibilités de reparation. Chambre des représentants de Belgique 19 novembre 2019. Doc 55 0771/001; Avis du conseil d'état n° 66.910/1 du 21 février 2020. Doc 55 0914/002).

Models that are, either fully or to some extent, based on a more flexible notion of "expected/normal product lifetime", have greater potential to be adapted to a modern approach that takes sustainability into account when assessing a product and its expected durability in any legal context. However, the problem is that this does not include any attempt to change the stance on "what is expected". Therefore, these models accept that, for instance, the expected cell phone lifespan is three to four years (see note 10) or, as demonstrated in Austrian research, that a consumer uses a t-shirt on average for 2.5 years, and a refrigerator for 9.4 years (Wieser, Troeger, and Hübner 2015). These kinds of models are also neutral due to the fact that, for instance, because of price competition, the average lifespan of a washing machine has dropped from ten years to under seven years in the last decade, and that the cheapest appliances last only a few years (Whitegoods Trade Association undated). Therefore, these kinds of norms reflect society "as is". If we want a transformation from linear models of consumption to circular models, these norms are not useful in their current forms.

A second example of normal or expected lifespan assessment is offered by intellectual property law. In situations where a patented product is modified without the patent holder's permission in an area of commercial activity, it must be assessed whether this is a "genuine" repair (which is permitted) or a reconstruction (which is considered as a patent infringement) as a part of so-called exhaustion doctrine.[12] While patent law is not fully harmonised in the EU and national variations exist between the EU Member States, usually the following aspects are assessed in drawing the line between a permitted repair and unpermitted reconstruction: to what extent the technical effects of the invention are embodied in the replaced component; what is the normal lifespan of the product as the product can be repaired only within that lifespan; what is the extent of the repair and does the repaired part compete with the original parts (see more, for instance, Pihlajarinne and Ballardini 2020: 244). The second criterion is interesting from the perspective of this article. An assessment of normal lifespan is based on whether the product has fulfilled its function: is its life at an end? The normal lifespan of a product has been assessed on the basis of the "common understanding in society".[13]

There is a potential problem from the CE perspective, since unsustainable products based on the linear model of consumption are usually considered as representing a "normal lifespan". The way the public perceives a product's lifespan is closely linked with the patent holder's explicit or implicit guidance on lifespan. There might be huge variations in terms of how the "normal lifespan" is seen by the public depending on the kind of choices made in the way of commercialising the patent, planning, and manufacturing of the product. The information given by the patent holder and seller in connection with marketing, for instance, explicitly about the product and its lifespan but also more implicitly in terms of guarantee period, utilisation possibilities, and future exchange value might serve as sources of input for what kind of expectations the public has.[14] On the part of rightsholders, there might be incentives to present a product's lifespan as shorter than it might potentially be (Pihlajarinne 2021: 92).

Other types of intellectual property rights – copyrights, designs and trademarks – might also serve as barriers for repairs. Exhaustion doctrine is usually also applied to recycling repair efforts targeted at products protected by these rights. Due to the fragmentation of intellectual property rights, some specific interests relate to each type of IPR, and due to the fact that international and EU harmonisation as well as national development efforts have mostly been focused on looking at one right at a time, they all are based on their own, detailed legislation. As a result, there are differences in the way that the exhaustion doctrine is applied: it is dependent on the right concerned as well as case-specific circumstances. While there is no such practice on explicit "normal lifespan" assessment as in patent law, a common principle applied to some extent to all those rights is that alterations changing the identity of the product are deemed infringements of intellectual property rights.[15] Moreover, the repair versus reconstruction dichotomy in patent law reflects the same principle: especially the "normal lifespan" assessment combined with the idea that the more extensive the action is, the more probable it is unpermitted. The "no alterations" principle together with "normal lifespan" assessment as its specific case follows the linear model of the notion of consumption: lifespan is seen as linear, where a product in its unchangeable form moves towards its unavoidable end. Therefore, copyright law, industrial design law and trademark law implicitly contain elements of "normal" or "expected" lifespan assessment similar to patent law.

A lifespan assessment conducted under the normative context of "normal" or "expected" lifespan under consumer law and intellectual property law is a potential tool for promoting circular product lifespans. However, in their current form, these norms reflect the linear model of production and consumption: lifecycles are perceived as a process where a product's life, without facing any interventions, terminates unavoidably after a relatively short period of use. It is a widely accepted legal standpoint that a product must correspond to the consumer's expectation. Consumers have a "right to normality", and legal norms take the throwaway culture "as is". The "normal lifespan" or "expected lifespan" requirement represents a passive legislator.

Consumer behaviour and perceptions of normality might be changing due to increasing calls for sustainability. EU policy will include new measures based on the new CE action plan (2020), such as improving product durability (widening the scope of the Ecodesign Directive) and the obligation to inform consumers on product lifespan. Even though these do not directly impact "expected" or "normal" lifespan evaluations, these kinds of actions might, in the long run, contribute positively to consumer attitudes. However, without legal intervention, the threshold of "normal" in private law regulation might transform only very slowly towards sustainable lifespan.

The "normal" or "expected" lifespan requirements represent a passive legislator: consumer expectations and behaviour patterns and the social and moral norms relating thereto dictate the lifespan threshold set by regulation. If linear production and consumption are seen as normality, then a transformation to CE requires that "abnormality" should be a target.

Could Nordic attitudes and legal culture be utilised as a new path towards a new normality?

As demonstrated above, Nordic consumer attitudes are in general positive towards CE-related actions. Another positive Nordic element might be Nordic legal culture. Nordic law is considered as an independent legal family, which has its own legal mentality.[16] Nordic law is close to civil law while it has only distant connections with common law traditions. However, a difference with civil law is a lack of extensive codification even though the position of statutes is high (Husa, Nuotio, and Pihlajamäki 2007: 11–12). However, the aspect that Nordic law shares with the common law tradition is the pragmatism of its legal culture and legal thinking when compared with the formalist and concept-centred approach of the civil law tradition. Systematisations and divisions are not as crucial as in systems based on the civil law tradition. By contrast, the focus is on concrete problems and feasible ways to tackle them (Husa, Nuotio, and Pihlajamäki 2007: 10; Husa 2010: 259–260; Schüller 2011: 387–388). Law is seen as a tool for "social engineering".[17] This means that the Nordic way of legal thinking emphasises more the impacts of regulation, and the aim of seeking feasible solutions instead of systematisation and categorisation of legal concepts as such. Even though precedents in practice enjoy high relevance, they lack the formally binding position that they enjoy in common law countries. Despite the fact that the courts have more room to decide compared with the civil law tradition (Husa 2010: 260–261), legal positivism and a strong position of parliament are reflected in a rejective attitude to court-made rules (Husa, Nuotio, and Pihlajamäki 2007: 9). Especially in Sweden and in Finland, courts are very cautious in matters relating to any kind of moral issues, which they are reluctant to assess without parliament's guidelines (Husa 2010: 257–260). While it is possible that the space reserved for courts' independent argumentation in Nordic legal culture is gradually widening (Husa 2010: 269), it is not probable that the Nordic courts will be eager for involvement in any kind of judicial activism in the future, either. However, if Nordic courts or corresponding authorities receive signals from the legislator, for instance in the form of committee reports and the like, they might dynamically adopt new ways of interpretation (see Pihlajarinne 2017). Therefore, even the legislator's "weak signals" (for instance, committee reports by the ministry) are important when law is faced with new societal needs.

Due to Nordic consumers' relatively dynamic attitude towards CE on the one hand and the pragmatism of their legal culture on the other, the Nordics could act as a forerunner in attempts at directing lifespan expectations towards more sustainable models. The sustainable lifespan requirement should be integrated into the thresholds of "normal" or "expected" lifespans in consumer and patent law. Since EU consumer law sets only a non-ambitious minimum requirement and national patent law is not harmonised, considerable space remains for national activity. The current situation is that the legislator leaves the lifespan assessment to "consumer behaviour reality", as current lifespan assessments are based on consumer expectations. The "normal" or "expected" lifespan requirements most probably reflect the social and moral norms that have over recent decades very much followed the linear model of consumption. This "consumer behaviour directs the requirement of normality in the regulation" idea should be complemented or accompanied by a contrastive method of input, where an active legislator, using regulation as a tool for social engineering, sets the requirement of "sustainable lifespan" as a new normality. This would shape consumer expectations, practices and in the longer run, also impact on the formulation of moral norms concerning lifespans.

If consumer law liability periods reflected the needs of a circular economy-based society, those liability periods should be detached from requirements based on short fixed-term liability periods or expected lifespan assessments. In that case, liability periods would not purely reflect the need to protect consumers' monetary and safety interests but also the societal interest of promoting consumers' possibilities to make sustainable choices. Consumer expectations in terms of sustainable lifespans might not yet be a reality except among the most environmentally conscious groups of consumers. However, the research discussed above indicates that even the average Nordic consumer's attitude might be more positive than elsewhere in Europe, so the "sustainable lifespan" requirement would expedite the ongoing transformation of Nordic consumers' attitudes in a more sustainable direction.

The best alternative for tackling the unsustainability problems relating to IPRs following the "normal lifespan" idea would be to reformulate exclusive rights so that they are limited within the sustainable lifespan idea. This would mean that the scope of infringing acts should be re-defined so that only acts that extend beyond the genuine purpose of maximising the lifespan of a product or material would be infringing acts. Another alternative would be to revise the "no alterations" principle by allowing alterations that are necessary for achieving a sustainable lifespan for the product category in question. In the patents and repairs context, a "sustainable lifespan" threshold for that particular category of product could be applied instead of the normal lifespan reflecting current consumer behaviour patterns. This change would mean a change from the idea of what the lifespan of a product is to an idea of what it should be, that is, how long the product should work in a CE-based society. (Pihlajarinne 2021: 97–99).

One major difficulty is how to assess a "sustainable lifespan". Prevailing perceptions on sustainable lifespan are under constant change, and great differences might exist between product categories. Courts should not be left alone to define sustainable lifespan but they should be provided with help from research-based assessments. Mechanisms for this should be developed.

Some of the proposed changes might need an explicit change in the law, such as abandoning short fixed-term liability periods. As copyright law, trademark law, and industrial design law are harmonised by the EU, the CJEU could develop a new doctrine on sustainable alterations in these areas of law. However, for instance, there should be nothing preventing the national courts from adopting new ways to interpret "expected" or "normal" lifespan requirements. However, considering, for instance, the Nordic courts' strong respect for the legislator's will, it might be necessary for the legislator to indicate that it would support applying a "sustainable lifespan" model instead of "normal or expected lifespan". A change might not be possible without institutional support from the legislator, if not legislative action for changing the law.

Conclusions

Consumer behaviour patterns and expectations towards the durability of products are of utmost importance in the transformation towards a circular economy. Research demonstrates that the Nordic countries' consumer attitudes are more positive in comparison with other European countries' consumer attitudes in terms of CE-related activities. However, these attitudes might not yet be realised in practice and behaviour patterns. This aspect, together with the Nordic pragmatic legal culture, could serve as fertile ground for Nordic legislators to be more ambitious and take an active role in extending product lifespans.

Private law, such as consumer law and IPR norms, applying a threshold of "expected" or "normal" lifespan for products indicates that the legislator takes consumer behaviour patterns as they are. From the CE perspective, this reflects a passive legislator letting current consumer expectations and behaviour patterns dictate the lifespan threshold required by regulation. These models indicate domination by linear models of consumption. Regulatory intervention in these norms would be feasible. The "consumer behaviour directs the requirement of normality in regulation" notion should be complemented/accompanied by a contrastive input method where an active legislator sets a "sustainable lifespan" requirement. This would shape consumers' expectations, practices and in the longer run, also impact on the formulation of moral norms concerning lifespans. This would be important especially from the perspective of a social and solidarity economy.

Notes

1 This article has been produced as part an Academy of Finland funded research project "Shaping, fixing and making markets via IPR: regulating sustainable innovation ecosystems" (SHARE) (332326). Excellent research assistance by Saija Partanen is gratefully acknowledged.

   For instance, Eléonore Maitre-Ekern (2019: 1, 78–79) defines the circular economy as "an economic model that is precautionary and allows the Earth system to regenerate and thrive, and to offer a safe operating space for humanity to prosper in the long term. It relies on renewable energy, aims to eliminate the use of hazardous substances, reduce resource and product consumption, and avoid final waste generation (notably through careful design for prolonged product lifetime and recovery), and focuses on social well-being".

2 Boulanger 2010: 5–6. About the differences between these strategies, see Maitre-Ekern 2021: 4.

3 Ghisellini, Cialani, and Ulgiati 2016: 11–32. Maitre-Ekern (2021: 2) recommends that EU policymakers take more extensive action on pre-market producer liability – which limits market access by minimum requirements for product durability, repairability, and reusability – along with an extended producer liability approach. She argues that an over-strong focus on waste management law and extended producer responsibility fails in dealing with environmental problems and that these approaches basically amount to attempts to fix the linear consumption model.

4 Dessein et al. (2015: 8) discuss the three roles that culture has in sustainability. One of them, "culture as sustainable development", indicates that all human decisions have cultural roots and therefore, culture is an important foundation for sustainable development.

5 The Charter of Fundamental Rights of the EU (Article 37) includes a similar provision: "A high level of environmental protection and the improvement of the quality of the environment must be integrated into the policies of the Union and ensured in accordance with the principle of sustainable development".

6 For instance, research by Henrik Riisgaard, Mette Mosgaard, and Kristina Overgaard Zacho (2016: 120–121) concluded that consumers are willing to use local repair shops and to pay for repair and maintenance services for cell phones in Denmark.

7 In this study, among the Nordic countries Finland was considered the most positive towards CE due to active CE-related discussion and stakeholder recommendations presented to the parliament and municipalities. The attitude of Finnish consumers was demonstrably the most positive among the Nordics for most CE-related action as well as for the possibility to reduce their consumption over the next five years (SB Insight 2019).

8 Naturally, beyond the scope of this chapter additional types of consumer law measures exist that do not directly relate to the liability period as such, but have an incentivising impact on lifespans. For instance, the French Consumer Code includes a prohibition on activities linked to planned obsolescence where a manufacturer's intention is demonstrably to "reduce the lifespan of the product".

9 The period was lengthened in Sweden because it was assumed that products will last more than two years (Regeringens proposition 2004/05:13 Distans- och hemförsäljningslag m.m. [Prop. 2004/05:13], 103).

10 In Norway, the five-year period is reserved for products or parts that can be expected to last significantly longer than two years under normal use. The reason for setting a system of two guarantee period categories is differences between product categories (Ot.pr p. nr. 44 [2001–2002] 158 Om lov om forbrukerkjøp [forbrukerkjøpsloven] Tilråding fra Justis- og politidepartementet av 15. mars 2002, godkjent i statsråd samme dag, 166). The Supreme Court of Norway has found (Norges Høyesterett – Dom HR-2007-1592-A, 21.9.2007) that the expected lifespan of a cell phone is three or four years. This was significantly longer than two years and so cell phones fall under the five-year guarantee period. The court stated that consumers' expectations were impacted by the information given at the moment of purchase that the cell phone industry had assessed the lifecycle of cell phones as three to five years.

11 HE Hallituksen esitys Eduskunnalle laiksi kuluttajansuojalain muuttamisesta ja eräiksi siihen liittyviksi laeiksi 360/1992 vp., 118. For instance, the Finnish consumer dispute board decided in case 5685/32/2015, 14.04.2018 that the consumer should be prepared to accept that after four-and-a-half years use of a dishwasher the need for maintenance services might emerge, which does not serve as an indication of a defect in the product. The same conclusion was drawn in case 2242/38/08, 20.01.2009 concerning a printer used over a two-year period.

12 The exhaustion doctrine limits the possibilities of an intellectual property holder to control a protected individual item after it has been released into the market with the rightholder's permission.

13 The German case: German Supreme Court, BGH, 17 July 2012, X ZR 97/11 (Palettenbehälter II) offers a good example. The court stated that in drawing the line between repair and reconstruction, it is important to assess whether the technical effect of a patented invention resides in the part exchanged. However, if consumers and trade circles believe that a replacement constitutes remanufacture of the patented product, the action constitutes a patent infringement, in spite of assessing whether the replacement reflects technical aspects of the innovation. Therefore, consumers' view on the perception of the product – assumptions of its use value as well as its exchange value – are relevant.

14 About problems relating to the patent holder's part, see also Heath and Môri 2006: 856.

15 This principle can be seen in the following cases, for instance: Case BGH 1-ZR 89/08 *Verlangarete Limousinen* serves as an example of its application in the industrial

designs context. A company had acquired Daimler cars and widened them by adding a section in the middle, and then put them on the market. The Bundesgerichtshof concluded that the exhaustion doctrine was not applicable because the product clearly deviated from standard version cars that the rightholder had put on the market. In the copyright context, the CJEU case *Art & Allposters International BV v. Stichting Pictoright*, C-419/13, the court stated that the applicability of the principle of exhaustion depends on "whether the altered object itself, taken as a whole, is physically the object that was placed onto the market with consent of the rightholder". Also in trademark law it is often argued that the principle of exhaustion might not be applied to situations where the identity of the original product has been turned into a new, independent product. In trademark law, a trademark holder who does not want product alterations can refer to the Trademark Directive, Article 15.2 which allows preventing further commercialisation of a product for a legitimate reason, "especially where the condition of the goods is changed or impaired after they have been put on the market".

16 Legal mentality means an overall attitude consisting of fundamental legal values, social fairness and social ethics (Husa 2010: 256).

17 Heikki Pihlajamäki (2004: 472) stresses that social engineering could be achieved in other ways than through legal realism, and that the emergence of legal realism in Scandinavia and the USA therefore has other explanations.

References

Bicchieri, Cristina. 2005. *The Grammar of Society: The Nature and Dynamics of Social Norms*. Cambridge, NY: Cambridge University Press.

Boulanger, Paul-Marie. 2010. "Three Strategies for Sustainable Consumption." *S.A.P.I.E.N.S* 3, no. 2., p. 1-10.

Camacho-Otero, Juana, Casper Boks, and Ida Nilstad Pettersen. 2018. "Consumption in the Circular Economy: A Literature Review." *Sustainability* 10, no. 8: 2758.

Dessein, J., Soini, K., Fairclough, G. and Horlings, L. (eds). 2015. "Culture in, for and as Sustainable Development." In *Conclusions from the COST Action IS1007 Investigating Cultural Sustainability*. University of Jyväskylä, Jyväskylä.

European Commission. 2011. *Attitudes of Europeans Towards Resource Efficiency: Analytical Report*. Flash EB Series no.316. Accessed June 2, 2021. https://europa.eu/eurobarometer/surveys/detail/902.

European Commission. 2018. *Behavioural Study on Consumers' Engagement in the Circular Economy - Final Report*. Prepared by LE Europe, VVA Europe, Ipsos, ConPolicy and Trinomics. Accessed June 2, 2021. https://ec.europa.eu/info/sites/info/files/ec_circular_economy_final_report_0.pdf.

Gallaud, Delphine, and Blandine Laperche. 2016. *Circular Economy, Industrial Ecology and Short Supply Chain. Vol. 4 of Innovation, Entrepreneurship, Management: Smart Innovation Set*. London: Wiley.

Ghisellini, Patrizia, Catia Cialani, and Sergio Ulgiati. 2016. "A Review on Circular Economy: The Expected Transition to a Balanced Interplay of Environmental and Economic Systems." *Journal of Cleaner Production* 114: 11–32.

Gnanapragasam, Alex, Masahiro Oguchi, Christine Cole, and Tim Cooper. 2017. "Consumer Expectations of Product Lifetimes Around the World: A Review of Global Research Findings and Methods." In PLATE: Conference Proceedings of PLATE 2017, 8–10 November 2017, Delft, the Netherlands, edited by Conny Bakker, and Ruth Mugge, 464–469. Amsterdam: IOS Press.

Guerin, Daniel, Jean Crete, and Jean Mercier. 2001. "A Multilevel Analysis of the Determinants of Recycling Behavior in the European Countries." *Social Science Research* 30, no. 2: 195–218. https://doi-org.libproxy.helsinki.fi/10.1006/ssre.2000.0694.

Heath, Christopher, and Mineko Môri. 2006. "Ending is Better than Mending: Recent Japanese Case Law on Repair, Refill and Recycling." *International Review of Intellectual Property and Competition Law* 37, no. 7: 856–864.

Howells, Geraint, Christian Twigg-Flesner, and Thomas Wilhelmsson. 2018. *Rethinking EU Consumer Law*. New York: Routledge.

Husa, Jaakko. 2010. "The Stories We Tell Ourselves: About the Nordic Law in Specific." In *Legal Culture and Legal Transplants: La culture juridique et l'acculturation du droit*, edited by Jorge A. Sánchez Cordero, 250–267. Washington, DC: International Academy of Comparative Law.

Husa, Jaakko, Kimmo Nuotio, and Heikki Pihlajamäki. 2007. "Nordic Law: Between Tradition and Dynamism." In *Nordic Law: Between Tradition and Dynamism*, edited by Jaakko Husa, Kimmo Nuotio, and Heikki Pihlajamäki, 1–39. Antwerpen: Intersentia.

Jackson, Tim. 2005. *Motivating Sustainable Consumption: A Review of Evidence on Consumer Behaviour and Behavioural Change*. Guildford, UK: Centre for Environmental Strategy, University of Surrey.

Kirchherr, Julian, Marko Hekkert, Ruben Bour, Anne Huijbrechtse-Truijens, Erica Kostense-Smit, and Jennifer Muller. 2017. "Breaking the Barriers to the Circular Economy." Deloitte. Accessed June 2, 2021. https://circulareconomy.europa.eu/platform/sites/default/files/171106_white_paper_breaking_the_barriers_to_the_circular_economy_white_paper_vweb-14021.pdf.

Maitre-Ekern, Eléonore. 2019. "Towards a Circular Economy for Products: A Legal Analysis of Europe's Policy and Regulatory Framework from an Ecological Perspective." PhD diss. University of Oslo.

Maitre-Ekern, Eléonore. 2021. "Re-Thinking Producer Responsibility for a Sustainable Circular Economy: From Extended Producer Responsibility to Pre-Market Producer Responsibility." *Journal of Cleaner Production* 286. https://papers.ssrn.com/sol3/papers.cfm?abstract_id=3732613.

Maitre-Ekern, Eléonore, and Carl Dalhammar. 2019. "Towards a Hierarchy of Consumption Behaviour in the Circular Economy." *Maastricht Journal of European and Comparative Law* 26, no. 3: 394–420. doi:10.1177/1023263X19840943.

Miliute-Plepiene, Jurate, Olle Hage, Andrius Plepys, and Algirdas Reipas. 2016. "What Motivates Households Recycling Behaviour in Recycling Schemes of Different Maturity? Lessons from Lithuania and Sweden." *Resources, Conservation and Recycling* 113: 40–52.

Oguchi, Masahiro, and Masaaki Fuse. 2015. "Regional and Longitudinal Estimation of Product Lifespan Distribution: A Case Study for Automobiles and a Simplified Estimation Method." *Environmental Science Technology* 49, no. 3: 1738–1743. https://doi-org.libproxy.helsinki.fi/10.1021/es505245q.

Pihlajamäki, Heikki. 2004. "Against Metaphysics in Law: The Historical Background of American and Scandinavian Legal Realism Compared." *American Journal of Comparative Law* 52, no. 2: 469–487.

Pihlajarinne, Taina. 2017. "Arkipäivän oikeusvertaileva näkökulma: yhdysvaltalaisista vaikutteista pohjoismaisessa immateriaalioikeudessa." *Lakimies* 155, no. 7–8: 1121–1140.

Pihlajarinne, Taina. 2021. "Repairing and Re-Using from an Exclusive Rights Perspective: Towards Sustainable Lifespan as Part of a New Normal?" In *IP and Sustainable Markets*, edited by Ole-Andreas Rognstad and Inger Berg Ørstavik. Edward Elgar Publishing, Cheltenham.

Pihlajarinne, Taina, and Rosa Ballardini. 2020. "Paving the Way for the Environment: Channelling 'Strong' Sustainability into the European IP System." *European Intellectual Property Review* 42, no. 4: 239–250.

Riisgaard, Henrik, Mette Mosgaard, and Kristina Overgaard Zacho. 2016. "Local Circles in a Circular Economy: The Case of Smartphone." *European Journal of Sustainable Development* 5, no. 1: 109–124.

SB Insight. 2019. *The Nordic Market for Circular Economy: Attitudes, Behaviours & Business Opportunities*. Accessed June 2, 2021. https://www.nordea.com/Images/33-308788/circular-economy-19-small.pdf.

Schlacke, Sabine, Klaus Tonner, and Erik Gawel. 2015. *Stärkung eines nachhaltigen Konsums im Bereich Produktnutzung durch Anpassungen im Zivil- und öffentlichen Recht*. Texte 72/2015. Umweltbundesamt. Accessed June 2, 2021. https://www.umweltbundesamt.de/sites/default/files/medien/378/publikationen/texte_72_2015_staerkung_eines_nachhaltigen_konsums_im_bereich_produktnutzung_0.pdf.

Schüller, Bastian. 2011. "Social Peace via Pragmatic Civil Rights: The Scandinavian Model of Consumer Law." In *The Many Concepts of Social Justice in European Private Law*, edited by Hans-W. Micklitz, 384–404. Cheltenham, UK: Edward Elgar Publishing.

Tonner, Klaus, and Rosalind Malcolm. 2017. *How an EU Lifespan Guarantee Model Could Be Implemented Across the European Union*. Directorate General for Internal Policies—Policy Department for Citizens' Rights and Constitutional Affairs. Accessed June 2, 2021. https://www.europarl.europa.eu/RegData/etudes/STUD/2017/583121/IPOL_STU(2017)583121_EN.pdf.

Whitegoods Trade Association. n.d. How Long Should It Last, accessed June 2, 2021, https://www.whitegoodstradeassociation.org/for-public-mainmenu-43/how-long-should-it-last-.

Wiesbrock, Anja, and Beate Sjåfjell. 2014. "The Importance of Article 11 TFEU for Regulating Business in the EU: Securing the very Basis of our Existence." In *The Greening of European Business under EU Law: Taking Article 11 TFEU Seriously*, edited by Beate Sjåfjell, and Anja Wiesbrock, 1–12. London: Routledge.

Wieser, Harald, Nina Troeger, and Renate Hübner. 2015. "The Consumers' Desired and Expected Product Lifetimes." In *PLATE 2015 Conference Papers*, Nottingham Trent University, UK, 17–19 June 2015. https://www.plateconference.org/consumers-desired-expected-product-lifetimes/.

9 Assessing through a gender-inclusion lens the social impact of circular strategies in the apparel value chain

The Dutch case

Lis Suarez-Visbal, Claudia Stuckrath, and Jesús Rosales Carreón

Introduction

The apparel value chain (AVC) comprises different industries, with diverse businesses operating in several geographical locations. It employs more than 9.3% of the global working population, ensuring the livelihood of millions of workers in the extraction, manufacturing, distribution, and end-of-life stages of its value chain (De Souza et al. 2010; Franco 2017; Ozturk et al. 2016; World Bank 2013). It is associated with increasing inequalities in how clothes are made, stressing the importance of evaluating its social impacts (Fletcher and Tham 2014; Seuring et al. 2008; Boström and Micheletti 2016; Franco 2017; Resta et al. 2016; Ellen MacArthur Foundation 2017; Shen et al. 2017). At the same time, the industry is highly feminised as more than 75% of workers are women occupying the lowest-paid jobs (Fletcher and Tham 2014), which raises crucial challenges regarding gender equality.

Businesses in the AVC have been adopting circular economy as a new framework to achieve sustainability, where environmental, economic, and social considerations are integrated into the business model (Henry et al. 2020). The circular economy can reduce environmental impacts by minimising resource flow and implementing different circular strategies such as resale, rental, repair, remanufacture, and recycling (Guldmann 2016; Stahel 2016; Jung and Jin 2016; Accenture 2019). So far, the economic and environmental dimensions have been successfully addressed by different businesses such as Nudie Jeans, Patagonia, Vigga, and Filippa K (Jung and Jin 2016; Pal et al. 2016; Ellen MacArthur Foundation 2017).

However, there is little knowledge about circular economy social impacts (e.g., decent pay, gender equality, labour conditions), and no known scientific framework to assess circular economy social impacts at the business level (Elia, Gnoni, and Tornese 2017; Merli, Preziosi, and Acampora 2018; Millar, McLaughlin, and Börger 2019). Recent assessment methods consider mainly environmental or economic domains (Corona et al., 2019). This vision often overlooks potential impacts and trade-offs such as regulation, governance,

DOI: 10.4324/9781003255246-9

culture, inclusivity, and marked inequalities, resulting in perspectives with incomplete information (Elia, Gnoni, and Tornese 2017; Geissdoerfer et al. 2017; Iacovidou et al. 2017; Merli, Preziosi, and Acampora 2018; Millar, McLaughlin, and Börger 2019).

Although several considerations of social impacts exist, circular economy's social dimension has been defined in literature mainly by the number of jobs created (Millar, McLaughlin, and Börger 2019). This definition is narrow in scope and depth as it does not define the type of job (or its quality), nor does it elaborate on potential individual and community impacts or potential trade-offs between different kinds of workers, which from a solidarity and inclusion point of view is also relevant.

Although there are numerous social impact assessment frameworks (SIAF) within the broader sustainability field and a myriad of sector-specific tools developed by NGOs and companies, these do not seem, on their own, to fully address the circular economy social impacts characteristics within the sector (Jijelava and Vanclay 2014; Vanclay 2002; Maa undated). Conventional linear activities in the AVC are characterised by i) labour intensity with low payment and low working conditions (Kane 2015; Ascoly 2009; Asia Floor Wage Alliance 2016), ii) the over-representation of vulnerable populations working at various stages of the AVC (Kate and Theuw 2016), and iii) the feminisation of the workforce (Van Nederveen Meerkerk 2018; Fletcher and Tham 2014). As circular strategies are being deployed in the sector, they risk adopting these low-pay low-working conditions for vulnerable workers.

Circular economy has intrinsic characteristics that could amplify effects in the sectors where it is implemented. First, it requires tight collaboration and well-defined reverse logistics with upstream and downstream stakeholders of the AVC to close the loops (Iacovidou, Hahladakis, and Purnell 2020). As it presupposes the development of technical innovations, it can create, displace, or eliminate jobs at various stages of the AVC. Thus it is pivotal to analyse circular strategies social impacts from a systemic level (Vanclay 2019). Second, with the implementation of these performance loops, existing low working conditions could be reinforced and create a lock-in situation (Kim 2000). Third, sorting and separating different materials in the recycling phase as well as preparing materials for remanufacturing can expose workers to unknown chemicals, with undetermined health and well-being issues for them and close by communities (Ellen MacArthur Foundation 2017; Pla-Julián and Guevara 2019; Ingallina 2018).

Therefore, strengthening the circular economy's social dimension is vital for theoretical and societal reasons. From a theoretical point of view, circular economy's claim as a new paradigm to achieve sustainability could be jeopardised if its social dimension is not reinforced. From a societal perspective, if the circular economy in the apparel sector weakly addresses the social dimension, jobs created could share the same prevalent low working conditions as in the linear business model. Consequently, workers in the sector will not be better off with circularity. Their weak and vulnerable position could be exacerbated if circularity in the sector increases in the future.

This chapter addresses the lack of a social impact assessment framework for circular economy with a gender-inclusion lens to assess the AVC. A gender-inclusion lens can help identify structural inequalities between men and women in socio-economic spheres and suggest a course of action to redress those inequalities.

We chose to test our proposed Social Impact Assessment Framework for Circular Economy with a Gender-Inclusion Lens (SIAF-CE♀) in the Dutch AVC because the Netherlands is a frontrunner in circular economy policy and implementation across various sectors (Kirchherr et al. 2018; Government of Netherlands 2016; Van Rompaey 2019).

The remaining segments are organised as follows: the second section explains how social impacts are considered in the circular economy concept. The third section describes the methodology. The fourth section presents our proposed social impact assessment framework. The fifth section discusses our results and delineates the most relevant topics for further analysis. Finally, the sixth section presents our conclusions.

Social impact assessment frameworks and circular strategies

Even though there is not currently a framework to assess circular economy social impacts, there are several SIAFs from different scales and-or levels of aggregation that could be used for this purpose. Most sector-specific tools engage collaboratively with stakeholders to either inform consumers or work specifically on workers' rights, capacity development, and voice of workers (Casey 2006). However, many of these frameworks are based on audits that have been highly criticised for relying on a compliance perspective (e.g., required by the brand) that leans too much on management and not so much on workers' feedback (Barrientos and Smith 2007).

According to Jijelava and Vanclay (2014), SIAFs should address social issues at three levels: the individual, household-community, and societal levels. Other important considerations are that i) job creation is how the current social ambition of circularity is defined, ii) quality, inclusivity, and community impacts are considered relevant aspects of job impacts so far unaddressed by circularity (Circle Economy 2020; Willeghems and Bachus 2018; Rubery 2019), and iii) the AVC is highly feminised. Understanding by feminisation the fact that most workers within the low-income and poor working conditions bracket are mostly women and that gender constructs such as lack of unequal access to education and family obligations prevent them from taking better opportunities and improving their livelihoods (Van Nederveen Meerkerk 2018; Fletcher and Tham 2014).

Given these considerations, we define three social impact dimensions that circular strategies in the AVC, should address which are detailed in the following sections.

Quality of jobs dimension

Although there is no one definition of quality of jobs (QOJ), there is consensus that it relates to the characteristics that affect employee work-life. Usually, it includes elements such as pay, working conditions, and career opportunities

(Burchell et al. 2014). Several frameworks focus on the QOJ, and we highlight three: the International Labour Organisation (ILO) work indicators, the United Nations Economic Commission for Europe (UNECE), and the OECD framework. The ILO decent work indicators compromise a set of 75 indicators and 21 legal frameworks. They cover ten main aspects that are summarised in Table 9.1. The ILO standard is often used as a reference point to help countries define the QOJ in a particular sector. However, because of its complexity, it is not practical for company-level applications.

The statistical framework developed by UNECE (Cazes, Hijzen, and Saint-Martin 2015) provides a set of indicators to measure employment quality with seven dimensions and 12 subdimensions. ILO and UNECE are very comprehensive and cover multiple dimensions that rely on numerous indicators of different nature, which from a business perspective is often seen as a problem as it affects the comparability and usability of these frameworks. None of them is based on normative choices about what should be considered as "good" or "bad" jobs, but rather, they guide how those indicators should be used (Cazes, Hijzen, and Saint-Martin 2015).

Finally, the OECD framework for job quality assessment (Cazes, Hijzen, and Saint-Martin 2015) is structured around three components: earning quality, labour market security, and work environment. It follows the guiding principles of the well-being agenda, as recommended in the Stiglitz Report (Stiglitz, Fitoussi, and Sen 2009). It focuses on both the job characteristics and job quality outcomes (as experienced by workers) and not only on drivers of job quality (such as regulation and compliance) (Cazes, Hijzen, and Saint-Martin 2015). "The job characteristics approach defines the quality of the working environment in terms of several specific characteristics that influence workers' wellbeing" (OECD 2017). This framework explicitly emphasises workers' perspectives as opposed to employers or investors.

Sustainable livelihood dimension

Sustainable livelihood (SL) refer to the living standards, assets, and opportunities that a household, family, or community enjoys (Cahn 2002). Regarding community well-being, two of the most widespread community poverty alleviation and well-being frameworks are the Multidimensional Poverty Index (MPI) developed by the Oxford Poverty and Human Development Initiative (Alkire et al. 2015) and the sustainable livelihood approach (DFID 2000). The MPI consists of ten indicators of poverty, divided into three broad dimensions. The MPI combines i) the proportion of poor people and ii) the intensity, defined as the percentage of dimensions in which the poor are divided (Alkire, Conconi, and Roche 2013; Alkire et al. 2015). The MPI is a comprehensive framework to assess poverty and community wellbeing at a national level. However, it does not have indicators to assess from a micro-perspective or organisational level.

The sustainable livelihood approach (DFID 2000) includes a portfolio of five assets (human assets, natural assets, physical assets, financial assets, and social assets) (Krantz 2001), out of which people construct their living. It is a flexible and

Table 9.1 Different aspects covered for the social impact assessment frameworks reviewed

		Compliance with laws and workplace regulations	Harassment or abuse	Compensation and benefits	Living wages	Over-time work	Free choice work	Health and safety	Social security	Skills training	Freedom of association and collective bargaining	Employment relations	Safe working conditions	Gender relations	Quality of life	Family and community impacts	Forced labour	Child labour	Discrimination	Cultural and community rights	Environment	Customs	Ethical business behaviour	Privacy and transparency
Cross-sector	Social Life Cycle Assessment (S-LCA)	●		●										●					●	●	●		●	●
	Social Impact Assessment (SIA)			●										●	●	●				●	●			
	Business Social Compliance Initiative (BSCI)			●		●		●			●						●	●	●		●		●	
Apparel-specific	Waste and Resources Action Programme (WRAP)	●	●	●		●	●	●	●		●								●		●	●		
	Fair Wear Foundation (FWF)	●	●	●	●	●	●	●			●						●	●	●					
	Fair Labour Association (FLA)	●	●	●		●	●	●			●						●	●	●					
	Ethical Trade Initiative (ETI)		●	●	●	●					●	●	●				●	●	●					
Specific to social impacts: QOJ, SL, and GE&I	Sustainable Livelihood (SL)			●	●	●				●					●	●				●	●			
	International Labour Organisation (ILO)			●		●		●		●	●	●	●		●	●	●	●						
	United Nations	●		●		●		●		●	●		●		●		●	●	●	●				
	Economic Commission for Europe (UNECE)			●											●									
	Organisation for Economic Co-operation and Development (OECD)										●		●	●				●						
	Gender Equality in Social Auditing guidance (BSR)		●	●							●		●	●					●					
	Eight Building Blocks from the International Center for Research on Women (ICRW)		●	●							●			●			●	●	●					

straightforward framework, adaptable to various projects, and it has been widely used in both developing and developed countries. However, it has been criticised for overseeing existing power dynamics, emphasising individuals' agency while not analysing structural barriers in-depth (Horsley et al., 2015; Scoones 2009).

Gender equality andiInclusion dimension

The gender equality and inclusion (GE&I) dimension is important because in assessing social impacts, there is a tendency to see the community as a homogeneous unit without considering different roles, positions, and situations of women and men (Vanclay 2002; Jijelava and Vanclay 2014). Such implicit assumptions result in knowledge of social impacts based only on one perspective (Van Nederveen Meerkerk 2018). With women and vulnerable populations disproportionately affected by adverse working conditions, it is vital to consider their perspective and the different roles regarding circular strategies implemented in the industry (Neetha 2002; Fletcher and Tham 2014).

There are several gender analysis tools made by academics, NGOs, and governmental agencies such as the OECD and the SIDA, the Capacities and Vulnerabilities Framework (CVA), the Equality and Empowerment Framework (Longwe), and the UN Gender Rapid Assessment Tool, among others (Equilo undated). Additionally, some apparel sector-based organisations have integrated a more comprehensive gender equality approach. Three available tools in this area are the eight building blocks for women economic empowerment (ICRW 2016), the BSR gender impact framework equality guide for social auditing and the data collection tool of the Social Convergence Project (SLCP undated). The BSR tool and SLCP are relatively novel initiatives with little widespread use by industry practitioners and no academic analysis. However, these tools share similar indicators and focus on the same areas as the eight building blocks for women's economic empowerment.

We argue that when analysing jobs and, in particular, circular jobs in the apparel sector, applying a transformative gender-inclusion lens can help identify structural inequalities between men and women in socio-economic spheres and suggest a course of action to redress those inequalities. Transformative is understood as a way that seeks a more systemic approach. By applying a gender lens, we analyse how economic and social development is determined by power relations in different spheres of society from the family, community and working relations. An inclusion lens highlights the blind spots where individuals with several vulnerabilities (e.g., gender, social class, ethnicity, place of origin, religion) are excluded from development processes. Social exclusion and inclusion are critical aspects of the social and economic policy agenda. According to Labonté and Hadi (2011: 3),

> although there is much disagreement among authors on what the concepts of social exclusion/inclusion mean, there is emergent consensus that they imply that (households and their members) have access to material resources, to

labour markets, education and healthcare, have freedom from discrimination, opportunities for social participation, and voice around the policy choices affecting all of these conditions.

In this sense, a transformative gender-inclusion lens should tackle four essential considerations: i) type of jobs and economic opportunity (English 2013; Van Nederveen Meerkerk 2018), ii) access to non-economic factors such as agency, empowerment, and autonomy (Kabeer 2013; Kabeer and Mahmud 2004), iii) intersectionality, defined as the interaction between race, class, and gender (Crenshaw 1991; Weldon 2006), and iv) existing power dynamics influencing social and job market discrimination rooted in existing socio-cultural structures (English 2013). This way, a gender-inclusion lens can provide valuable insight to inform targeted businesses and policymaking recommendations (Harcourt 2019; Weldon 2006).

Methods

With the considerations mentioned above and based on these three dimensions explained, we built the SIAF-CE♀. We used a mixed-methods approach explained in seven steps based on Bell and Morse's (2008) *imagine strategy to develop sustainability indicators and frameworks.*

Step 1 – contextualisation (data collection I)

Through desk research and semi-structured interviews, we gather the social impacts and social impact assessment considerations relevant to the AVC. Our sample consisted of two groups: direct businesses implementing circular strategies and indirect ecosystem stakeholders (i.e., think tanks, academics, policymakers, NGOs, and workers' rights organisations). We created two different sets of interviews, one for each group. Business interviews were used to validate the potential social impacts of circularity and their social impact assessment considerations. Indirect stakeholders' interviews were set up to corroborate the context of the circular strategies. Using a snowball sampling method, we interviewed 25 businesses and five indirect stakeholders. All interviews were recorded, transcribed, and coded thematically. Given that interviews were anonymous, a code was given to each stakeholder.

Step 2 – creation and application of boundary criteria

As existing frameworks and indicators are both a benchmark and comparison to national and international standards, we privilege this approach to create our boundary criteria (Miller, Buys, and Summerville 2007). We reviewed 203 documents (143 scientific articles, five theses, and 55 governmental, NGO, and independent think-tank reports). Keywords such as: "social impact and circularity", "circular strategies in the apparel industry", "social impact of circular strategies", and "social impacts indicators in the apparel value chain" were used. Following this logic, we identified 40 SIAFs from existing literature and interviews, and we reviewed them according to the following criteria:

Cr1: It applies to the apparel sector.
Cr2: It is used by both practitioners and academics in the field.
Cr3: It has been used in developed and developing countries.
Cr4: It covers at least one of the three identified social dimensions of circular strategy (QOJ, SL, GE&I).

Step 3 – creation and application of practicality criteria

Practical issues are one of the main reasons a SIAF fails to be applied by businesses (Kendall and Knapp 2000; Barman 2007; Nicholls 2009). Therefore, based on our content analysis (step 1) and considering Mahmoudi et al. (2013), Veleva and Ellenbecker (2001), and Miller et al. (2007), we constructed a set of practical criteria that were applied to the SIAFs identified after step 2. The practicality criteria are:

Cr5: It is flexible, adaptable, and business-friendly.
Cr6: It is comparable among different countries, types of businesses, and circular strategies.
Cr7: It considers workers' perspectives and the specific gender and workers aspects of circular strategies in the AVC (highly feminised and heavy employment of vulnerable populations with precarious working conditions).

Step 4 – final selection of SIAFs

To the remaining frameworks after step 3, we performed a strength and weaknesses analysis to decide which frameworks were the most suitable for assessing the social impacts of circular strategies in the AVC. Three frameworks and eight building blocks were combined into the SIAF-CE♀ to reduce the individual framework's shortcomings identified through our desk research.

Step 5 – operationalisation

To operationalise the SIAF-CE♀, we defined dimensions, categories, and indicators for each concept included in the SIAF-CE♀. We also determined questions and scores for each indicator. However, when existing frameworks included questions associated with each indicator, those questions were taken and adapted to fit our research constraints when needed. (Set of corresponding SIAF-CE♀ questionary available upon request.)

We normalised the selected frameworks to the same scale and score system. A worker's survey with five open-ended questions and 85 closed-end questions was developed based on this information. This survey was the main tool to collect data, along with semi-structured interviews. All preselected frameworks suggested Likert scale and Likert-type scale questions with variating points from four to seven. We defined a four-point Likert scale (to avoid a neutral answer).

Composite indicators were calculated for each answer using the average of the values. If the result was not an integer number, it was normalised by approximating the closest integer. In this way, we converted the values into the same qualitative scale.

Step 6 – framework validation

The proposed SIAF-CE♀ was validated with the 3S-Methodology, from Cloquell-Ballester et al. (2006). Following this method, conceptual coherence, operational coherence, and utility of indicators and framework were validated first by the researcher, second by academia and experts, and third by practitioners. Conceptual coherence defines the correct relation between the measuring instrument (indicator/ framework) and the measuring object (environmental/social quality). Operational coherence determines the correct definition of the measuring instrument's internal operations (indicator/framework). Last, utility determines the applicability of the indicators/framework in environmental and social assessment studies (Cloquell-Ballester et al., 2006). Conceptual coherence was validated using the extent to which a dimension/category/indicator assesses a relevant part of circular strategies in the AVC. Operational coherence and utility were validated using both the extent to which a dimension/category/indicator applies to companies and data availability. The validation was based on surveys sent to nine validators and the relative level of consensus using a four-point Likert scale, where one was little (relevance/applicability/availability), and four was very relevant.

Step 7 – testing of the SIAF-CE♀ (data collection II)

The Dutch AVC was used as a case to test our SIAF-CE♀. The target population for our surveys consisted of male and female workers from Dutch businesses in the AVC currently adopting circular strategies. The sampling method was clustering, and 60 workers surveys were conducted. Surveys were confidential in-person and due to COVID-19 measures also via online meetings. The open-ended questions of the surveys were analysed following a thematic coding to set up the context as explained in step 1. The closed-ended questions were analysed employing a four-point Likert scale using descriptive statistics (i.e., mean, standard deviation, mode, and frequency). Workers' surveys were tested on reliability through Cronbach's alpha analysis.

Results and discussion

Relevant social impacts and social impact assessment considerations for circularity

Based on our desk research and corroborated with the interviews, in general, social impacts present in the AVC should also be considered in circular strategies, including at least safety and health, living wages, gender equality, excess overtime, and voice and agency. According to (EXP2): "Even though we have been talking and doing sustainability in apparel for almost 20 years, most of these issues are still not solved. So, in this transition to circularity, many of those social impacts will continue to be an issue".

Even though 82% of interviewees acknowledged the importance of the social dimension on their business, only 53% measured it, showing the need for more accessible and practical assessment mechanisms that can be used at various stages of the AVC. There is also the need to make the social impact ambition of circular economy more explicit from the business and policymakers perspective.

Four reasons explain why businesses have a low social impact assessment. First, there is an asymmetry in how the industry values social impacts and environmental impacts (Norman and MacDonald 2004). Companies tend to see environmental impacts as a cost-saving strategy. In contrast, they see social impact as a cost. As (GOV1) expressed: "When companies prioritise risk, they have a due-diligence process in mind, which generally has a higher focus on environmental measures, and some are seeing Circular Economy as a way to reduce some of those risks". Second, social impacts are perceived as harder to assess than environmental impacts (Kendall and Knapp 2000; Barman 2007; Nicholls 2009), which was echoed by several interviewed businesses. Third, there is scepticism in the field around assessment tools and their effectiveness (Barrientos and Smith 2007), as confirmed by many enterprises interviewed. As put by (NGO2): "Social impacts are measured by what you want to avoid (the base of current social audits). To be transformative, we also need to include what you want to see happening". The fourth reason is that there is no current framework to analyse circularity's social impacts. This lack of a proper framework to assess social impacts keeps relevant aspects hidden (Saidani et al. 2019; Millar, McLaughlin, and Börger 2019; Murray, Skene, and Haynes 2017). As (S-CS1) indicated: "I have not seen a particularly universal framework around social impact measuring in the Circular Economy".

Chosen frameworks to assess social impacts of the circular economy

After applying our boundary criteria, a total of 13 assessment tools were selected for further review. Table 9.1 summarises the different SIAFs reviewed and the social aspects that these frameworks addressed.

Table 9.2 illustrates the evaluation of the selected frameworks after applying the practicality criteria. Columns C1 to C7 show the fulfilment (●) of a criterion for a specific framework. Only the frameworks that fulfilled all seven criteria were kept and underwent a subsequent strength and weakness analysis. The OECD quality of job framework (Cazes, Hijzen, and Saint-Martin 2015), the SL framework (DFID 2000), the eight building blocks for women's economic empowerment (ICRW 2016), and the Gender Equality in Social Auditing Guidance (Barraja, 2019) were chosen to fulfil our proposed three-dimensional SIAF-CE♀. The OECD quality of job framework (Cazes, Hijzen, and Saint-Martin 2015) was chosen because it provides historical and geographical data from developed and

Table 9.2 Selection process of the social impact assessment frameworks chosen including the fulfilment of criteria and strength and weakness analysis

Social impact assessment framework	Cr1	Cr2	Cr3	Cr4	Cr5	Cr6	Cr7	Strengths	Weaknesses	Final selection
Waste and Resources Action Programme (WRAP)	•	•	•	•						
Fair Wear Foundation (FWF)	•	•	•	•						
Fair Labour Association (FLA)	•	•	•	•						
Ethical Trade Initiative (ETI)	•	•	•	•						
Social Life Cycle Assessment (S-LCA)	•	•	•	•						
Social Impact Assessment (SIA)	•	•	•	•						
United Nations Economic Commission for Europe (UNECE)						•				
International Labour Organisation (ILO)	•	•	•	•						
Multidimensional Poverty Index (MPI)	•	•	•	•						
Sustainable Livelihood (SL)	•	•	•	•	•		•	Adaptable to individual, household, community	Not deep structural barrier	•
Organisation for Economic Co-operation and Development (OECD)	•	•	•	•	•		•	Workers focus, comparable, global, accessible	Not gender focused	•
Gender Equality in Social Auditing Guidance (BSR)	•	•	•	•	•		•	Strong gender base targeted to field complexities	Need a more complex power analysis	•
Eight Building Blocks from the International Center for Research on Women (ICRW)	•	•	•	•	•		•	Set of must gender equality principles, business-friendly	Not a framework just a set of principles	•

Criterion used

Cr1: Applies to the AVC

Cr2: Used by practitioners and academics

Cr3: Implemented in both developed and developing countries

Cr4: Covers at least one of social impacts (QOI, SL, or GE&I)

Cr5: Flexible, adaptable, and business-friendly

Cr6: Comparable among different geographies, types of businesses, and circular strategies

Cr7: Considers workers perspective and the specific gender and workers characteristics of circular strategies in the AVC

developing nations, facilitating comparisons. It also defines job quality from the workers' well-being perspective. The SL (DFID 2000) was chosen because it is adaptable to assess individuals, households, and community levels. The eight building blocks for women's economic empowerment (ICRW 2016) was chosen because it highlights essential areas to address when analysing gender equality and inclusion in a paid-work context. We also chose the BSR Gender Equality in Social Auditing Guidance (Barraja, 2019) as it incorporates notions of intersectionality and socio-cultural context. It also has been tested by more than 30 global companies.

Operationalisation of the SIAF-CE♀

Following Miller et al. (2007), the final SIAF-CE♀ was constructed by combining and adapting a selected set of four existing frameworks and building blocks. As a result, the SIAF-CE♀ has three dimensions, QOJ, SL, and GE&I. It comprises 15 composite, multi-attribute, qualitative indicators (Table 9.3). It uses 90 question worker's surveys, divided into four sections to assess each dimension and the socio-demographic context. A four-point Likert scale is used to score indicators for each dimension. The qualitative scale adopted was 1) low, 2) medium-low, 3) medium-high, and 4) high. Results of surveys are plotted on an excel database with their average for each indicator.

The QOJ dimension is based on 16 questions from the OECD framework to assess job quality. In the OECD QOJ framework, we adapted the earning quality indicator from gross hourly wage in purchasing power parity (PPP) to real monthly gross salary. This adaptation was made to better reflect the context of the AVC, where work schedules tend to favour part-time work (ETUI 2012), contributing to precarious working conditions. Additionally, to facilitate comparability among countries, the gross salary was transformed into the four-point Likert scale adjusted as follow: below or equal to the poverty line (low = 1), between the poverty line and minimum wage (medium-low = 2), between minimum wage and living wage (medium-high = 3) and above living wage and average salary on the sector (high = 4). Wages were converted into Euros PPP and compared using living wages data. Also, given that distribution of sector-specific earning indicator data on a national level was not available, this indicator was not used. The other two indicators of job security and work environment were kept as proposed by the OECD framework. The SL dimension has five composite indicators that represent each of the five assets of sustainable livelihoods. It was constructed based on 18 attributes from the UNDP SL sustainable development guide (UNDP 2017). The GE&I dimension is composed of seven indicators. For this dimension, nine attributes were developed with the BSR Gender Equality Social Auditing Guidance (Barraja, 2019) along with the eight building blocks and our literature review.

A flower allegory is used to illustrate the framework (Figure 9.1). Each of the three social dimensions is characterised by a layer of petals, representing an indicator for that dimension. The QOJ dimension comprises three indicators, represented by the most central layer of petals, and provides an individual perspective

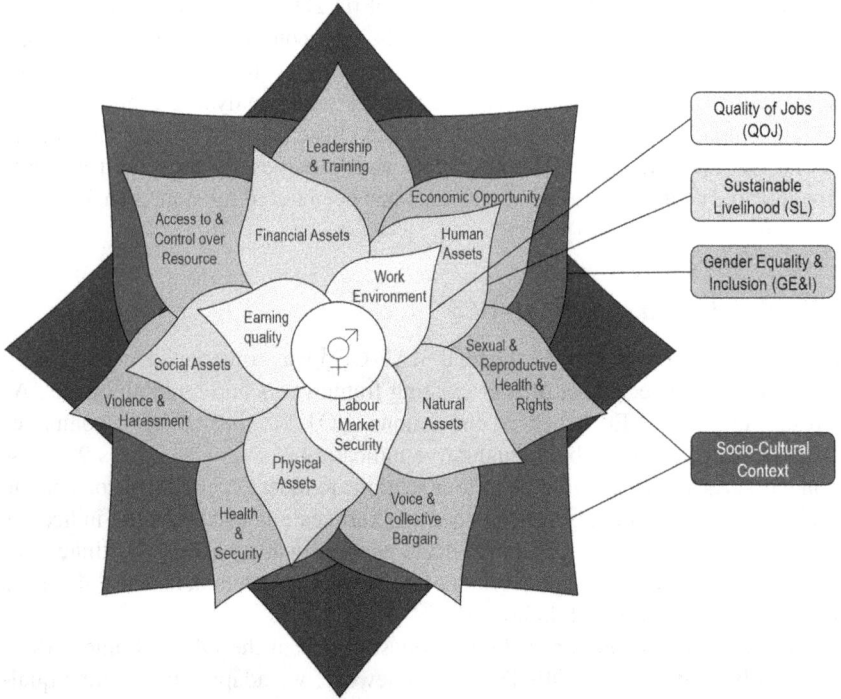

Figure 9.1 The flower SIAF-CE♀ model. Each layer of petals represents one social dimension, and each petal represents an indicator. Outside layers represent socio-cultural context and power dynamics.

on job characteristics. The SL dimension assesses workers' household and community well-being through sustainable livelihoods. Five petals-indicators represent it. The outer seven petals layer represents the GE&I dimension, which assesses gender equality and inclusion within the context of the job, the family, and the societal level. The outer layer represents the socio-cultural context and the power dynamics that need to be analysed within.

Validation of the SIAF-CE♀

All three dimensions were validated according to the 3S method (see third section). According to the validators, the three dimensions were considered relevant and applicable to the context of circularity within the Dutch AVC. In terms of data availability, only two indicators were thought to be easily available with a score above 80% (earning quality and health and safety), while five other indicators scored just over 50% making them potentially available. Indicators within the SL dimension showed the lowest availability score below 50% (human assets, personal assets, and social assets), which indicates that availability for these indicators is low. From the GE&I indicators, freedom from violence and freedom of movement also have a low availability score.

Anticipating this hard-to-get data issue, we designed both our framework's questionnaire and interview setting in such a way to make workers feel safer to talk, improving our data availability. This design included four elements: i) pre-testing questions with similar workers iterative in each location until hard questions were adequately addressed by tone, intensity, or language used, ii) conducting surveys in their chosen language, by gender-sensitivity trained staff, iii) adapting to inter-viewees availability (nights, weekends, online, personal way), and iv) including company-management in the process (especially human resources and corporate social management) and highlight the virtues of having an in-house social-impact assessment tool.

The SIAF-CE visual representation

To facilitate comparison and communication of the results, we created a visual representation that consists of a circular bar chart showing the value of the indi-cators from 1) low (no bar) to 4) high (outer limit) (Figure 9.2). The graph is divided equally into three areas to plot the indicators of each dimension (QOJ, SL, and GE&I). Each of these areas contains its corresponding indicators. Each of the indicators is shown with two bars, one bar for females and another for male workers.

The value illustrated for each indicator corresponds to the average without normalisation into the qualitative scale. It is a standard good practice that helps visualise and compare results, but it requires more interpretation on the user's part (Wilson et al. 2015).

Social implications of circular strategies in the Dutch context

To test the framework, the SIAF-CE♀ was applied to the Dutch sector. According to our interviewees, the transition to circularity in the Dutch AVC is happen-ing organically amid a stagnating economic context in the "end-of-life" textile subsector. Thus, creating additional jobs is not seen as an immediate priority. According to (I-TC1): "The global used textile market is compressing, and the quality of textile collected decreases". Additionally, according to (EXP1): "In the Netherlands, most of the jobs related to circular strategies gravitate around subsi-dised companies whose main mission is to employ populations that are at-a-dis-tance from the job market (including immigrants, refugees and other vulnerable groups)". This (particularly) seems to be the case for the local remanufacturing, resale, and recycling (collecting, sorting, and recycling) strategies. In companies where these strategies are implemented, workers often earn the minimum wage.

Our findings from the desk research and interviews show that, even within the traditional AVC in the Netherlands, the retail sector employs most people (Papú Carrone et al. 2021), circular strategies such as repair, resale and rental seem to create few quality jobs. As said by (NLS07), "repair is an extension of

Table 9.3 The attributes and description of the SIAF-CEↄ indicators

Dimensions	Indicators	Attributes	Description
Quality of jobs (QOJ)	Earning quality	Gross income per month	Gross income per month, calculated at exact hours work
		Distribution of earnings with all workers (earning inequality)	N/A
	Labour market security	Risk of unemployment (risk and expected duration)	Perceived risk of losing the job because of contract or conditions
		Unemployment insurance (the coverage of the benefits and their generosity)	Perceived access to social welfare if employment is lost
	Work environment	Time pressure at work	Job demands for working under stress, at high speed or tight deadlines
		Physical health factors	Exposure to chemicals, or tiring working positions and perceived level of support from colleagues
		Autonomy and learning opportunities	Skill training opportunities, opportunities for promotion or career advancement
		Workplace relationship	Perception of relationships with colleagues and supervisors and value of work
		Working very long hours	Exposure to unusual working schedules and flexibility at work
Sustainable livelihood (SL)	Human assets	Level of perceived health	Perception of health level
		Ability to work and retained work	Incidence of unemployment in the last two years
		Level of education and or skills training of household members	The highest level of education obtained
	Natural assets	Access to natural resources	Access and use of land, forest, water, and clean air
		Environmental quality	Access to good waste and recycling services
			Perceived level of cleanness and maintenance of community surrounding
	Physical assets	Access to good housing	Access to basic amenities in the household
		Access to transportation services and proximity of food sources	Access to public transportation
		Access to child/elderly care or recreation facilities	Access to childcare facilities or services
			Access to recreational facilities and playgrounds
	Social assets	Support from family and friends	Access to quality family time
			Perceived closeness with neighbourhood and community
		Participation in community life	Use of programs offered in the community
			Member of community or environmental or political group
	Financial assets	Income/savings	Capacity to pay all monthly expenses at ease
			Capacity to save monthly
		Money management (debts)	Level of debts (incapacity to save)
		Possession of goods	Ownership of assets on a household level

Gender equality and inclusion (GE&I)	Economic opportunity	Access to equitable, safe and reliable employment	Perception of women having equal access to job opportunities, internal promotion, and salary than men
	Access to and control over resource	Access to resource	Access to economic and financial assets
		Control over resources	Control and decision over own earnings
	Leadership and training	Access to training and development of leadership skills	Perception of women having equal access to training and leadership opportunities promoted by management
	Voice and collective bargain	Voice and collective bargain	Level of participation in unions or workers committee
	Violence and harassment	Procedure knowledge	Awareness of policy and procedure for addressing violence and harassment in the workplace
		Risk of violence	Perceived level of safety in the different physical spaces of workplace
		Norms and culture	Level of acceptance of victims of violence and harassment
	Health and security	Access to healthcare	Access to health systems and protective gear
		Risk of accidents	Prevention and treatment of accidents and injuries
	Sexual and reproductive health and rights	Sexual and reproductive health and rights	Sharing responsibility for childcare and housework

the customer service experience", and they are not a principal income-generating activity. These strategies tend to be considered tasks of current sales staff, creating jobs shifts rather than new jobs. Only new startups rental and resale platforms, for which these circular strategies are core business, can potentially generate jobs. However, in the current state of development, they rely heavily on volunteer jobs. Additionally, as validated by our workers' surveys, circular strategies in the sector are characterised by a predominance of temporary contracts and part-time work schemes, which is no different from general conditions in the retail sector in the Netherlands.

Figure 9.2a shows the social impact in the Dutch AVC with the SIAF-CE♂ presented in the radar bar chart visual. The chart is divided into three sections to show each of the dimensions, and within each dimension are graphed the results of each indicator disaggregated by male and female workers. Figure 9.2b shows a layer disaggregated by each circular strategy. This visual representation gives a snapshot view of circular strategies' social impact in the AVC using a gender-inclusion lens. Different layers could be added to show disaggregated data to allow a deeper analysis.

In general, the results show that women have lower positive social impacts than men (score lower than men). For instance, while the quality of job indicators is valued in the medium-high scale for men, it is medium-low for women workers, as they are the primary earners of minimum wage. Even though the Dutch minimum wage is considered by some a living wage, given that the Netherlands has a comprehensive social security system, the wage is still in the lower spectrum in the sector. More so, it seems that within different circular strategies, women are also being disproportionally represented in this wage bracket, as in the traditional AVC, while they are also the holders of most part-time jobs (Fabo and Belli 2017). However, altogether the QOJ indicators combined with labour security and work environment is medium-high. Further research should investigate comparisons among different geographies with different social security schemes.

When assessing each individual circular strategy, we can see contrasting realities between them, in particular regarding the QOJ indicators (earning quality and the GE&I indicators). The earning indicator is the lowest for rental startup workers and highest in the sorting recycling centre. Voice and collective bargain are also very different in rental (the lowest) and resale (the highest), which can be attributed to the different types of enterprise and the state of consolidation of these businesses. It should be the subject of further research.

In the SL dimension, the social assets are the lowest indicator related to the social network, family and neighbourhood related relations. Respondents' results show little connection with the neighbourhood or their immediate community. These elements relate to inclusion and should be looked deeper into in future research. Financial assets are also low and lower for women, which means they have little capacity to save as their income is just enough to live on. Many make minimum wage or just over, which is also a characteristic of the feminisation trajectory of the linear AVC.

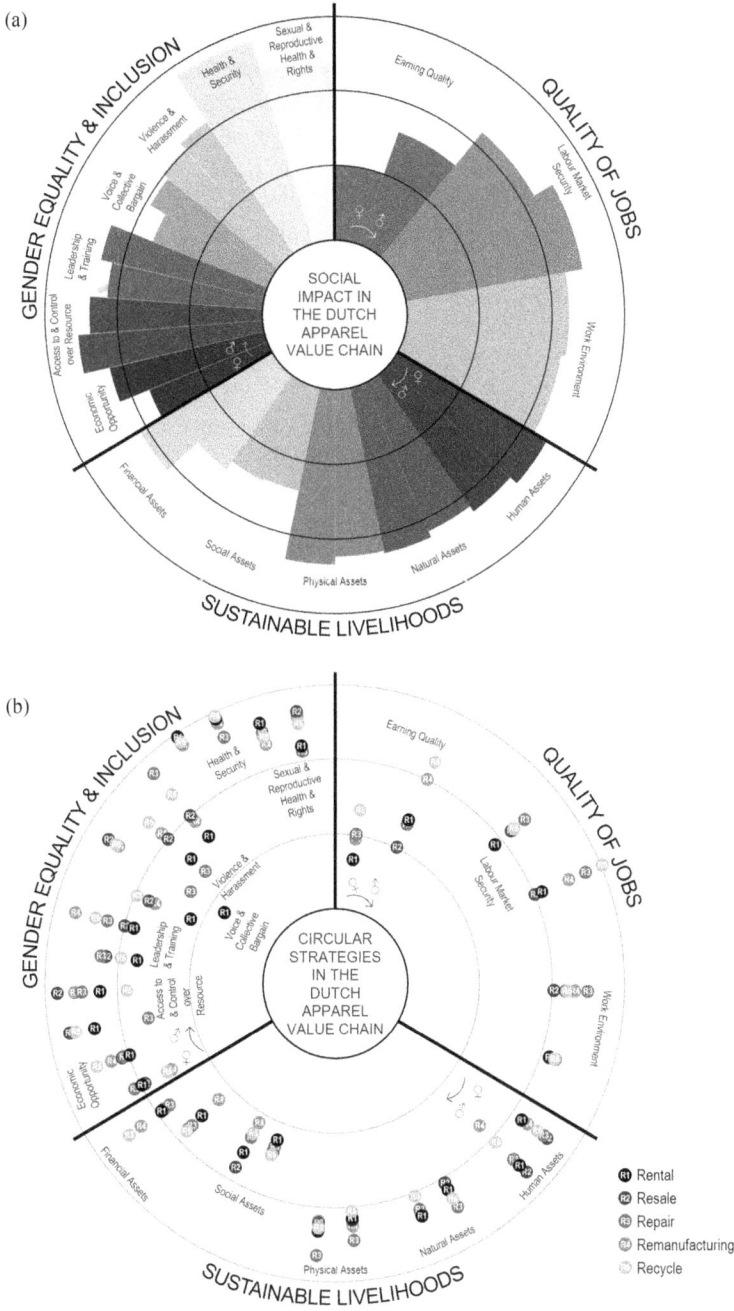

Figure 9.2 a) Visual bar chart representation of the social impact of circular strategies in the Dutch apparel value chain by gender; b) visual bar chart representation disaggregated for each circular strategy, represented by a specific R-number.

Finally, concerning the GE&I dimension, women perceive that they have lower access to economic opportunities, promotion, leadership, and training opportunities than men, but equal control of economic resources as men. Men, however, do not perceive any difference between themselves and women in this regard, which might point to a lack of awareness of inequality issues in corporate settings, also, for circular strategies. Regarding violence and harassment, it seems that male and female workers do not consider this indicator an issue. However, as seen in (Figure 9.2b), workers have contrasting perceptions among different circular strategies around this indicator. This outcome could also point to the difficulty of discussing violence and harassment openly. Future research should further elaborate into understanding how present these issues are within circular strategies.

Conclusion

This chapter addressed the lack of a framework for social impact assessment for the circular economy in the apparel value chain while showing its application in the Dutch context. We propose an assessment framework called the Social Impact Assessment Framework for Circular Economy with a Gender-Inclusion Lens (SIAF-CE♀) in the AVC.

Our results in the Dutch case indicate that local circular strategies show evidence of a differential in the impact on male and female workers, with positive but mitigated quality of jobs, sustainable livelihood, and gender equality and inclusion. Applying a gender-inclusion gives a more detailed perspective on how different workers are affected and helps to avoid reproducing existent patterns within an AVC that adopts circular strategies.

When comparing our SIAF-CE results with our literature review and interviews around working conditions in the AVC, our findings suggest that circularity in the Dutch AVC is currently not transformational as it seems to emulate the low working conditions and patterns of the linear model in the sector. At this moment, workers in the sector are not necessarily better off with circularity as the feminisation trajectory found in the traditional AVC persists even when circular strategies are adopted.

The proposed SIAF-CE♀ has several advantages; first, by combining existing tools and frameworks, we address the shortcomings identified in the literature of these tools individually. Second, our framework provides gender-disaggregated data highlighting the worker's perspective and the impact that jobs can have on their well-being and that of their family. Third, it is made to be used in multiple locations and provide an aggregated result for each circular strategy or by country or by company.

Lastly, the SIAF-CE♂ provides a multidimensional view of social impacts relevant to the AVC. It also provides a tool and a baseline to track the social impacts of circular strategies as they are being implemented. This approach can minimise potential trade-offs while offering businesses and policymakers a complete picture when assessing circular economy employment conditions' gendered socio-economic effects at the business level. This is pivotal to strengthening the take-up of circular strategies as an alternative to the take-make-dispose model. This SIAF-CE♀ could be relevant to other sectors that share similar characteristics of feminisation, labour intensity, and multiple locations as the AVC.

References

Accenture, Fashion for Good. 2019. 'The Future of Circular Fashion: Assessing the Viability of Circular Business Models'. https://reports.fashionforgood.com/wp-content/uploads/2020/10/The-Future-of-Circular-Fashion-Report-Fashion-for-Good.pdf

Alkire, Sabina, Adriana Conconi, and Jm Roche. 2013. 'Multidimensional Poverty Index 2013: Brief Methodological Note and Results'. *Oxford Poverty and Human Development Initiative* Oxford Department of International Development, University of Oxford *(OPHI)*. https://doi.org/10.13140/RG.2.1.4665.3049.

Alkire, Sabina, James Foster, Suman Seth, Maria Emma Santos, Jose Manuel Roche, and Paola Ballon. 2015. 'Multidimensional Poverty Measurement and Analysis: Chapter 1 Introduction'. *SSRN Electronic Journal*. https://doi.org/10.2139/ssrn.2564702.

Ascoly, Nina. 2009. 'The Global Garment Industry and the Informal Economy: Critical Issues for Labour Rights Advocates'. *Clean Clothes Campaign*, April 2009. http://www.cleanclothes.org/resources/publications/04-09-informal-labour-seminar-discussion-paper-ccc.pdf.

Asia Floor Wage Alliance. 2016. 'Precarious Work in the H&M Global Value Chain'. https://asia.floorwage.org/wp-content/uploads/2019/10/Asia-Floor-Wage-Alliance-H-M.pdf.

Barman, Emily. 2007. 'An Institutional Approach to Donor Control: From Dyadic Ties to a Field-Level Analysis'. *American Journal of Sociology* 112 (5). https://doi.org/10.1086/511802.

Barrientos, Stephanie, and Sally Smith. 2007. 'Do Workers Benefit from Ethical Trade? Assessing Codes of Labour Practice in Global Production Systems'. *Third World Quarterly* 28 (4): 713–29. https://doi.org/10.1080/01436590701336580.

Bell, S., and S. Morse. 2008. *Sustainability Indicators: Measuring the Immeasurable?* (2nd Ed.). Routledge. https://doi.org/10.4324/9781849772723

Boström, Magnus, and Michele Micheletti. 2016. 'Introducing the Sustainability Challenge of Textiles and Clothing'. *Journal of Consumer Policy* 39: 367–75. https://doi.org/10.1007/s10603-016-9336-6.

Barraja, M. 2019. *Gender Equality in Social Auditing Guidance*. BSR

Burchell, Brendan, Kirsten Sehnbruch, Agnieszka Piasna, and Nurjk Agloni. 2014. 'The Quality of Employment and Decent Work: Definitions, Methodologies, and Ongoing Debates'. *Cambridge Journal of Economics* 38 (2): 459–77. https://doi.org/10.1093/cje/bet067.

Cahn, M. 2002. 'Sustainable Livelihoods Approach: Concept and Practice'. In DevNet Conference 2002: Contesting Development: Pathways to Better Practice.

Casey, Roseann. 2006. 'Meaningful Change Raising the Bar in Supply Chain Workplace Standards'. Cambridge, Ma.

Cazes, Sandrine, Alexander Hijzen, and Anne Saint-Martin. 2015. *Measuring and Assessing Job Quality: The OECD Job Quality Framework*. OECD Social, Employment and Migration Working Papers no.174. https://doi.org/10.1787/1815199X.

Circle Economy. 2020. 'The Social Economy: A Means for Inclusive & Decent Work in the Circular Economy?' https://www.circle-economy.com/resources/the-social-economy-a-means-for-inclusive-decent-work-in-the-circular-economy.

Cloquell-Ballester, V. A., R. Monterde-Díaz, and M. C. Santamarina-Siurana 2006. 'Indicators Validation for the Improvement of Environmental and Social Impact Quantitative Assessment'. *Environmental Impact Assessment Review* 26 (1): 79–105. https://doi.org/10.1016/j.eiar.2005.06.002.

Corona, Blanca, Lin Shen, Denis Reike, Jesus Rosales-Carreon, and Ernst Worrell. 2019. 'Towards Sustainable Development through the Circular Economy: A Review and Critical Assessment on Current Circularity Metrics'. *Resources, Conservation and Recycling* 151. (104498): 1–15. https://doi.org/10.1016/j.resconrec.2019.104498.

Crenshaw, Kimberle. 1991. 'Mapping the Margins: Intersectionality, Identity Politics, and Violence against Women of Color'. *Stanford Law Review* 43 (6): 1241–99. https://doi.org/10.2307/1229039.

DFID. 2000. 'Sustainable Livelihoods Approach and Its Framework'. *Development*.

Elia, Valerio, Maria Grazia Gnoni, and Fabiana Tornese. 2017. 'Measuring Circular Economy Strategies through Index Methods: A Critical Analysis'. *Journal of Cleaner Production* 142: 2741–51. https://doi.org/10.1016/j.jclepro.2016.10.196.

Ellen MacArthur Foundation. 2017. 'A New Textiles Economy: Redesigning Fashion's Future'. *Ellen MacArthur Foundation.* https://www.ellenmacarthurfoundation.org/publications/a-new-textiles-economy-redesigning-fashions-future.

English, Beth. 2013. 'Global Women's Work: Historical Perspectives on the Textile and Garment Industries'. *Journal of International Affairs* 67 (1): 67–82.

Equilo. n.d. 'Gender Analysis'. Accessed 30th of may 2020. https://www.equilo.io/tools

ETUI. 2012. *Discounting the Workers: Conditions in the Retail Sector.* European Trade Union Institute.

Fabo, Brian, and Sharon Sarah Belli. 2017. '(Un)Believeable Wages? An Analysis of Minimum Wage Policies in Europe from a Living Wage Perspective'. *IZA Journal of Labour Policy* 6 (1). https://doi.org/10.1186/s40173-017-0091-3.

Fletcher, Kate, and Mathilda Tham. 2014. *Routledge Handbook of Sustainability and Fashion.* London: Routledge. https://doi.org/10.4324/9780203519943.

Franco, Maria A. 2017. 'Circular Economy at the Micro Level: A Dynamic View of Incumbents' Struggles and Challenges in the Textile Industry'. *Journal of Cleaner Production* 168: 833–45. https://doi.org/10.1016/j.jclepro.2017.09.056.

Geissdoerfer, Martin, Paulo Savaget, Nancy M.P. Bocken, and Erik Jan Hultink. 2017. 'The Circular Economy: A New Sustainability Paradigm?' *Journal of Cleaner Production* 143: 757–68. https://doi.org/10.1016/j.jclepro.2016.12.048.

Government of Netherlands. 2016. 'A Circular Economy in the Netherlands by 2050', 1–72.

Guldmann, Eva. 2016. *Best Practice: Examples of Circular Business Models.* The Danish Environmental Protection Agency. 1st ed. Copenhagen, Denmark: Danish Environmental Protection Agency/ Miljøstyrelsen. https://doi.org/10.13140/RG.2.2.33980.95360.

Harcourt, Wendy. 2019. 'What a Gender Lens Brings to Development Studies'. In *Building Development Studies for the New Millennium.* EADI Global Development Series 361–80. Cham: Palgrave Macmillan.

Henry, Marvin, Thomas Bauwens, Marko Hekkert, and Julian Kirchherr. 2020. 'A Typology of Circular Start-Ups: Analysis of 128 Circular Business Models'. *Journal of Cleaner Production* 245 (1). https://doi.org/10.1016/j.jclepro.2019.118528.

Horsley, Julia, Sarah Prout, Matthew Tonts, and Saleem H. Ali. 2015. 'Sustainable Livelihoods and Indicators for Regional Development in Mining Economies'. *Extractive Industries and Society* 2 (2): 368–80. https://doi.org/10.1016/j.exis.2014.12.001.

Iacovidou, Eleni, Joel Millward-Hopkins, Jonathan Busch, Philip Purnell, Costas A. Velis, John N. Hahladakis, Oliver Zwirner, and Andrew Brown. 2017. 'A Pathway to Circular Economy: Developing a Conceptual Framework for Complex Value Assessment of Resources Recovered from Waste'. *Journal of Cleaner Production* 168 (1): 1279–88. https://doi.org/10.1016/j.jclepro.2017.09.002.

Iacovidou, Eleni, John N. Hahladakis, and Phil Purnell. 2020. 'A Systems Thinking Approach to Understanding the Challenges of Achieving the Circular Economy'. *Environmental Science and Pollution Research.* 28: 24785–24806. https://doi.org/10.1007/s11356-020-11725-9

ICRW. 2016. *The Business Case for Women 's Economic Empowerment : An Integrated Approach.*

Ingallina, Stefano. 2018. *The Social-Economic Impact of Circular Economy in Europe: A Comparative Analysis*, 1–73.

Jijelava, David, and Frank Vanclay. 2014. 'Social Licence to Operate through a Gender Lens: The Challenges of Including Women's Interests in Development Assistance Projects'. *Impact Assessment and Project Appraisal* 32 (4): 283–93. https://doi.org/10.1080/14615517.2014.933505.

Jung, S., and B. T Jin. 2016. 'The Approach, Sustainable Development of Slow Fashion Businesses: Customer Value.' *Sustainability* 8: 540.

Kabeer, Naila. 2013. 'Paid Work, Women's Empowerment and Inclusive Growth. Transforming the Structures of Constraint'.

Kabeer, Naila, and Simeen Mahmud. 2004. 'Globalisation, Gender and Poverty: Bangladeshi Women Workers in Export and Local Markets'. *Journal of International Development* 16 (1): 93–109. https://doi.org/10.1002/jid.1065.

Kane, Gillian. 2015. 'Facts on India's Garment Industry'. *Clean Clothes Campaign* 2 (November 2014): 1–13. https://www.cleanclothes.org/resources/publications/factsheets/cambodia-factsheet-february-2015.pdf.

Kate, Gisela ten, and Martje Theuw. 2016. 'Fact Sheet Migrant Labour in the Textile and Garment Industry'. Centre for Research on Multinational Corporations -SOMO, no. February: 1–11. https://www.somo.nl/nl/wp-content/uploads/sites/2/2016/02/migrant-labour-in-the-textile-and-garment-industry.pdf.

Kendall, Jeremy, and Martin Knapp. 2000. 'Measuring the Performance of Voluntary Organisations'. *Public Management: An International Journal of Research and Theory* 2 (1): 105–32. https://doi.org/10.1080/14719030000000006.

Kim, Daniel H. 2000. *Systems Thinking Tools: A User's Reference Guide.* https://thesystemsthinker.com/wp-content/uploads/2016/03/Systems-Thinking-Tools-TRST01E.pdf.

Kirchherr, Julian, Laura Piscicelli, Ruben Bour, Erica Kostense-Smit, Jennifer Muller, Anne Huibrechtse-Truijens, and Marko Hekkert. 2018. 'Barriers to the Circular Economy: Evidence From the European Union (EU)'. *Ecological Economics* 150: 264–72. https://doi.org/10.1016/j.ecolecon.2018.04.028.

Krantz, L. 2001. *The Sustainable Livelihood Approach to Poverty Reduction.* SIDA. Division for Policy and Socio-Economic Analysis.

Labonté, R. N., A. Hadi, and X. E. Kauffmann. 2011. 'Indicators of Social Exclusion and Inclusion: A Critical and Comparative Analysis of the Literature'. *Working Papers* 2 (8). https://www.researchgate.net/profile/Detlef-Brem/post/anyone-know-of-a-theory-of-study-on-staff-members-perception-of-an-organizations-language-policy-and-the-potential-feelings-of-inclusion-exclusion/attachment/59d63903c49f478072ea5c0a/AS%3A273708856348672%401442268685177/download/Labonte+et+al+Indicators+of+Social+inclusion+and+exclusion+2012.pdf.

Maa, Karen. n.d. *Social Impact Measurements, towards a Guideline for Managers*. Rotterdam.

Mahmoudi, Hossein, Ortwin Renn, Frank Vanclay, Volker Hoffmann, and Ezatollah Karami. 2013. 'A Framework for Combining Social Impact Assessment and Risk Assessment'. *Environmental Impact Assessment Review* 43 (December 2017): 1–8. https://doi.org/10.1016/j.eiar.2013.05.003.

Merli, Roberto, Michele Preziosi, and Alessia Acampora. 2018. 'How Do Scholars Approach the Circular Economy? A Systematic Literature Review'. *Journal of Cleaner Production* 178: 703–22. https://doi.org/10.1016/j.jclepro.2017.12.112.

Millar, Neal, Eoin McLaughlin, and Tobias Börger. 2019. 'The Circular Economy: Swings and Roundabouts?' *Ecological Economics* 158: 11–19. https://doi.org/10.1016/j.ecolecon.2018.12.012.

Miller, Evonne, Laurie Buys, and Jennifer A Summerville. 2007. 'Quantifying the Social Dimension of Triple Bottom Line: Development of a Framework and Indicators to Assess the Social Impact of Organisations'. *International Journal of Governance & Business Ethics* 3 (3): 223–37. https://doi.org/10.1111/j.1467-8683.2007.00590.x.

Murray, Alan, Keith Skene, and Kathryn Haynes. 2017. 'The Circular Economy: An Interdisciplinary Exploration of the Concept and Application in a Global Context'. *Journal of Business Ethics* 140 (3): 369–80. https://doi.org/10.1007/s10551-015-2693-2.

Nederveen Meerkerk, Elise Van. 2018. 'Women Workers of the World United A Comparative History of Households, Gender and Work'. Inaugural lecture, Radboud University Nijmegen, 21 December 2018.

Neetha, N. 2002. 'Flexible Production, Feminisation and Disorganisation: Evidence from Tiruppur Knitwear Industry'. *Economic and Political Weekly* 37 (21): 2045–52.

Nicholls, Alex. 2009. '"We Do Good Things, Don't We?": "Blended Value Accounting" in Social Entrepreneurship'. *Accounting, Organisations and Society* 34: 755–769. https://doi.org/10.1016/j.aos.2009.04.008.

Norman, Wayne, and Chris MacDonald. 2004. 'Getting to the Bottom of "Triple Bottom Line"'. *Business Ethics Quarterly* 14 (2): 243–62. https://doi.org/10.5840/beq200414211.

OECD. 2017. 'OECD Guidelines on Measuring the Quality of the Working Environment'. Paris, France.

Ozturk, Emrah, Hasan Koseoglu, Mustafa Karaboyaci, Nevzat O. Yigit, Ulku Yetis, and Mehmet Kitis. 2016. 'Sustainable Textile Production: Cleaner Production Assessment/ Eco-Efficiency Analysis Study in a Textile Mill'. *Journal of Cleaner Production* 138 (2): 248–63. https://doi.org/10.1016/j.jclepro.2016.02.071.

Pal, Rudrajeet, Jonas Larsson, Heikki Mattila, and Markku Honkala. 2016. 'Local Fashion Value Chains: Success Factors and Competitive Advantages'. In *90th Textile Institute World Conference, 25-28 April*, 631–38. Poznan, Poland.

Papú Carrone, Natalia, Luis Sosa Lagunes, Hilde Van Duijn, Jade Wilting, Julie Metta, Tim Goesaert, and Kris Bachus. 2021. *Putting Circular Textiles to Work: The Employment Potential of Circular Clothing in the Netherlands*. Edited by Laxmi Haigh and Ana Birliga Sutherland. Circle Economy, HIVA.

Pla-Julián, Isabel, and Sandra Guevara. 2019. 'Is Circular Economy the Key to Transitioning towards Sustainable Development? Challenges from the Perspective of Care Ethics'. *Futures* 105 (September 2018): 67–77. https://doi.org/10.1016/j.futures.2018.09.001.

Resta, Barbara, Paolo Gaiardelli, Roberto Pinto, and Stefano Dotti. 2016. 'Enhancing Environmental Management in the Textile Sector: An Organisational-Life Cycle Assessment Approach'. *Journal of Cleaner Production* 135 (1): 620–32. https://doi.org/10.1016/j.jclepro.2016.06.135.

Rompaey, Stefan Van. 2019. 'C&A Makes Progress towards Sustainable Clothing'. *Fashion*, June 2019. https://www.retaildetail.eu/en/news/fashion/ca-makes-progress -towards-sustainable-clothing.

Rubery, Jill. 2019. 'A Gender Lens on the Future of Work'. *Journal of International Affairs* 72 (1): 91–106. https://jia.sipa.columbia.edu/gender-lens-future-work.

Saidani, Michael, Bernard Yannou, Yann Leroy, François Cluzel, and Alissa Kendall. 2019. 'A Taxonomy of Circular Economy Indicators'. *Journal of Cleaner Production*. https://doi.org/10.1016/j.jclepro.2018.10.014.

Scoones, Ian. 2009. 'Livelihoods Perspectives and Rural Development'. *Journal of Peasant Studies* 36 (1): 171–96. https://doi.org/10.1080/03066150902820503.

Seuring, Stefan, Joseph Sarkis, Martin Müller, and Purba Rao. 2008. 'Sustainability and Supply Chain Management - An Introduction to the Special Issue'. *Journal of Cleaner Production* 16 (15): 1545–51. https://doi.org/10.1016/j.jclepro.2008.02.002.

Shen, Bin, Qingying Li, Ciwei Dong, and Patsy Perry. 2017. 'Sustainability Issues in Textile and Apparel Supply Chains'. *Sustainability* 9: 1592. https://doi.org/10.3390/su9091592.

SLCP. n.d. 'Social and Labour Convergence'. Accessed 20 March 2020. https:// slconvergence.org/about-us/.

Souza, Antônio Augusto Ulson De, Aline Resmini Melo, Fernando Luiz Pellegrini Pessoa, and Selene Maria De Arruda Guelli Ulson De Souza. 2010. 'The Modified Water Source Diagram Method Applied to Reuse of Textile Industry Continuous Washing Water'. *Resources, Conservation and Recycling* 54: 1405–11. https://doi.org/10.1016/j .resconrec.2010.06.001.

Stahel, Walter R. 2016. 'The Circular Economy'. *Nature* 137–151. Springer, Cham. https:// doi.org/10.1038/531435a.

Stiglitz, Joseph, Jean-Paul Fitoussi, and Amartya Sen. 2009. 'Rapport de La Commission Sur La Mesure Des Performances Économiques et Du Progrès Social (Extraits)'. *Regards Sur l'actualité*. www.stiglitz-sen-fitoussi.fr.

UNDP. 2017. 'Application of the Sustainable Livelihood Model UNDP'. Panama City, Panama.

Vanclay, Frank. 2002. 'Conceptualising Social Impacts'. *Environmental Impact Assessment Review* 22 (3): 183–211. https://doi.org/10.1016/S0195-9255(01)00105-6.

———. 2019. 'Reflections on Social Impact Assessment in the 21st Century'. *Taylor & Francis Online*, 2019. https://doi.org/10.1080/14615517.2019.1685807

Veleva, Vesela, and Michael Ellenbecker. 2001. 'Indicators of Sustainable Production: Framework and Methodology'. *Journal of Cleaner Production* 9 (6): 519–49. https:// doi.org/10.1016/S0959-6526(01)00010-5.

Weldon, S. Laurel. 2006. 'The Structure of Intersectionality: A Comparative Politics of Gender'. *Politics and Gender* 2 (2): 235–48. https://doi.org/10.1017/S1743923X06231040.

Willeghems, Gwen, and Kris Bachus. 2018. *Employment Impact of the Transition to a Circular Economy: Literature Study*. Brussels.

Wilson, David C., Ljiljana Rodic, Michael J. Cowing, Costas A. Velis, Andrew D. Whiteman, Anne Scheinberg, Recaredo Vilches, Darragh Masterson, Joachim Stretz, and Barbara Oelz. 2015. '"Wasteaware" Benchmark Indicators for Integrated Sustainable Waste Management in Cities'. *Waste Management* 35: 329–42. https://doi .org/10.1016/j.wasman.2014.10.006.

World Bank. 2013. *World Development Report 2013: Jobs*. Washington, DC. https:// openknowledge.worldbank.org/handle/10986/11843.

10 From global problem to local solution

How a future-directed circular economy can foster social change

Sabine Lettmann and Pia Schmoeckel

Introduction

The world today is not sustainable, not resilient, and not fair for the majority of mankind. Human activity, and more in particular economic activity, pushes the planet beyond its natural boundaries. Science and technology demonstrate that the consequences of the existing economic, social, and environmental challenges become threatening to (human) existence on this planet and to the earth as we know it. Environmental exploitation and climate change directly contribute to an increase in global migration movements (Klepp 2017) resulting in societal restructuring. Accompanied by a growing number of people being disadvantaged in local societies, not at least influenced through a drifting income gap, this development emphasises a need for more balanced local structures in addition to tackling ecological but also economic challenges. Society is confronted with "parallel discourses" (Redclift 2005) on economic growth, social development, and environmental protection, (Daly 2007: 12) resulting in a potential clash between the interests and needs of the present generations and the ability of future generations to meet their needs (WCED 1987: 43; Mebratu 1998). One way to overcome the potential dichotomy between present and future generations and the tension between the discourses is provided by "sustainable entrepreneurship" by seizing the opportunities for simultaneously balancing "economic health, social equity and environmental resilience" (Cohen and Winn 2007). Economic health and environmental resilience are, for example, promoted through the system of a circular economy. The circular economy has gained momentum since the 1970s as a future-forward economic model being defined as restorative and regenerative by design. In general, it aims to keep products, components, and materials at their highest utility and value at all times. As the Ellen MacArthur Foundation (2015) states:

> The concept [...] is a continuous positive development cycle that preserves and enhances natural capital, optimises resource yields, and minimises system risks by managing finite stocks and renewable flows. It works effectively at every scale.

DOI: 10.4324/9781003255246-10

In contrast to other economic models, a circular economy focuses on redefining a growing economy "of which a wider society benefits" (Ellen MacArthur Foundation undated). Considering the latter definition and in particular the lack of clarification of the meaning behind the term "benefit", from a systematic perspective changing the economy from linear to circular structures as a sustainable business approach primarily addresses material flows. In more detail, the circular economy has been criticised as being "silent on the social dimension" with no explicit mentioning of social perspectives within conceptual approaches for sustainable development (Murray et al. 2017). However, according to Apostolopoulos et al. (2018: 2) the potential impact of entrepreneurial activity is not necessarily limited to environmental aspects. Moreover, they state entrepreneurship can further serve as a powerful driver for social change through maximising the impact of the Sustainable Development Goals such as reducing poverty (Apostolopoulos et al. 2018: 3). Small and medium enterprises (SMEs) represent 98% of all European businesses and contribute to 50% of the European GDP (European Commission 2020). Presumably, they hold an immense opportunity to drive social change combined with circular strategies when embedded into mainstream business activities. Therefore, it needs to be asked how a circular economy can recognise social equity and cultural inclusivity as a proactive response to level out societal challenges and moreover, how brands can be supported in merging circularity and social activism as part of their business operations. This aligns with Sheperd and Patzelt (2011) who define entrepreneurship as a "vehicle for empowerment" which, according to Holt and Littlewood (2018), may be generated throughout value chains.

As the process of design is not specified in the definition of a circular economy per se, this paper draws from terms such as design activism and social design to investigate ways fostering a shift to sustainable economic systems intertwined with an active formation of multicultural and inclusive communities. *Design activism* in the context of this case study refers to the definition of design by Victor Margolin (2015: 24) who concludes that for many people design has come to mean a process of envisioning an activity – rather than a product – that leads to a specific outcome which is useful to many. He further specifies that to generate a new social vision, a "Good Society", the connection of local, national and international organisations of designers with active citizens worldwide is required. Following the definition of Armstrong et al. (2014) the term *social design* comprises concepts and activities within participatory approaches to researching and generating new structures. Instead of predominantly following commercial objectives, these aim to make change happen towards collective and social ends. Additionally, Markussen (2017) refers to further criteria complementing the idea of social design and social value with aim, modus operandi, locus of design and innovation, and the scale of effects. In this paper, the definition of social design combines both theories to highlight a more comprehensive approach aiming to actively shape individual inclusion as well as to impact local communities. Whilst the activity of design relates to the latter definitions, its purpose as a driver for social change is framed according to Stephan et al. (2016). They identify positive social change as a transformation of mindsets, behaviour, and structural elements enabling beneficial outcomes for individuals, society, and/or the environment which are greater than benefits for the instigators themselves. Stephan et al.'s findings consider four main focus areas namely environment, social and economic inclusion, health and well-being, and civic engagement which may involve several domains simultaneously. To summarise, this paper looks at the circular economy and sustainable entrepreneurship as complementary, synchronised (Pedersen et al. 2018) business forms aiming at driving social change through unique incorporation of design activism and/or social design. Identifying these, this case study explores five entrepreneurial projects across Europe, including the Netherlands, UK, Germany, Italy, and Denmark predominantly located within different consumer goods sectors. They were selected as typical examples for working with refugees, disabled people or migrants whilst focusing on craftsmanship as an underpinned process of social change can be classified as social enterprises with an additional environmental focus expressed through the application of circular business strategies.

The first case, Arbor, evolved from a student initiative as part of Enactus, a global student-led NGO and is run by a master student in Sustainability Science and Policy since 2018. Arbor creates notebooks from post-consumer waste reusing single-sided printed paper mainly collected from University printers in Maastricht, Netherlands. The recycled notebooks are produced together with autistic men at a local social organisation whilst incorporating Maastricht Art Academy's test prints or packaging waste as individual front covers. The second, Bridge & Tunnel, is a small enterprise founded in 2016 in Hamburg, Germany. Under the leadership of a designer and a cultural scientist, this project focuses on upcycling pre- and post-consumer waste jeans to design fashion, bags, and other accessories. By working with six immigrants and long-term unemployed people from their local community, they integrate socially disadvantaged individuals being less or inexperienced workers into their team of 11 people in total. Deanderezeep is another project from the Netherlands creating certified sustainable soaps together with refugees. They pay attention to keeping project support and participant numbers even and their team currently comprises 15 people. Soaps are wrapped in post-consumer waste material and sold at the weekly farmers' market connecting students, refugees, and the local community. Deanderezeep belongs to the student-led social entrepreneurship association Enactus Maastricht. The next project, Mending for Good, is a project that offers creative, design-driven solutions to luxury brands for the issues of waste and excess stock. The repurposing of dead stock is done specifically through a network of high-quality craft projects linked to social cooperatives based in Italy. Its value-driven solutions are mindful of the environment and deliver positive social impact, while supporting the circular philosophy. The proposal to re-make, re-purpose, re-craft brands' excess raw materials, creates virtuous collaborative partnerships. This project initiates working opportunities and awareness for craft as an integrative process. Finally, the last participant in this case study is Sheworks Atelier which evolved from a research project through a collaboration with the municipality in Kolding and the Designskolen Kolding in Denmark. Led by designer and founder Solveig Søndergaard, 15 women from different cultural backgrounds work together in a team to create textile interior products primarily from pre-consumer waste sourced from Danish textile companies. The nexus of design thinking and problem-solving creates jobs for the women involved while fabric waste is reduced. Eight women are employed in the design studio which is now an independent company.

Although projects differ in size and product focus, they mutually aim to create jobs for their participants whilst reducing pre- or post-consumer waste. Their solution-focused response to unemployment, social disadvantages, and waste is locally connected and context-dependent besides being driven by design thinking and multidimensional learning.

Methodology

To investigate general project frameworks, benefits, and challenges from working within multi-cultural settings, online interviews were conducted via Zoom with each of the five project leaders. Interviews were semi-structured with predetermined open-ended questions allowing detailed responses and unexpected insights whilst increasing reliability and consistency among all interviews conducted. Responses were recorded and transcribed for evaluation giving qualitative information about each individual project whilst a coding process allowed to identify main themes (Harding and Whitehead 2015: 129). Additionally, project leaders completed an online survey comprising twelve different statements to cover the subjects discussed in interviews. Participants were asked to relate their answers to a scale from 0 (low) to 5 (high) aiming at categorising their experiences. Thus, quantifying the collected data enabled a comparison between projects even though diverse in subject, experience and structure and with this to develop a clearer picture of shared values, challenges, and project impacts. Likewise, it supported finding variations and nuances within the information collected. To keep project statements anonymous, naming in the results section has been replaced by "Project A–E" and does not correlate to the order outlined in the previous project introduction above. Throughout the interview and the raw data analysis data security of the participants' responses remained.

Case study

In the first part, this study seeks to find out how each of the five investigated project's business structure, product development, and manufacturing supports local, social engagement. Furthermore, it aims to identify on which level a collective mindset embracing equality enhances how project leaders and participants acquire skills as a form of social inclusion beyond business, instigated through a bidirectional learning environment. Hence, selecting projects from different European countries offered a broader multicultural perspective on similar enterprise structures embedded into various cultural environments. Interviews with project leaders highlight similarities in the strengths of a social circular economy as well as obstacles limiting the scaling of production and business resilience. By demonstrating the latter in the second part, this research further aspires to guidance for upcoming initiatives to facilitate successful frameworks regardless of their geographical location to allow social change to become the heart of circular business models instead of being a side effect only.

Results

Business structures

All projects involved in this research align their business policies with two main areas: firstly, with ways to positively impact environmental issues such as waste reduction, for example, through applying strategies based on the pillars of the circular economy: to design out waste and pollution, keep products and materials in use, and to regenerate natural systems (Ellen MacArthur Foundation 2015). Secondly, they align with the principles of corporate social responsibility (CSR) identified by Hazarika (2013) as "a company's integral responsibility towards its primary stakeholders including employees, customers, investors and suppliers, and […] the company's increased social improvements". Projects develop consumer goods ranging from accessories to garments and utilise their processes to create social value "as part of an activity" (Choi et al. 2018). All projects started on a small scale but have grown since then. While one project already sells internationally, some have opened online shops or developed B2B collaborations with smaller retailers. Others sell their products directly on local markets which they see as a further way of cultural exchange when project participants are connected with local communities. From the beginning, all brands had a mutual vision to socially innovate in terms of Moulaert et al. (2013) "driven by aspirations to increase social inclusion and well-being through improving social relations and empowerment processes". They all combine environmental and social aspects ranging from believing in social justice, inclusion and offering workplaces for people who are disadvantaged in a regular labour market due to a lack of qualification, language barrier, or migration status. Nuances in aspirations range from "finding a common language other than verbal" (Project A) to "increasing problem-solving and multi-dimensional learning through design thinking" (Project E). According to Julier and Kimbell (2019), an organisational process that investigates and generates solutions often goes way beyond the boundaries of experiences of products and services.

Waste material utilisation as the second core element all projects have in common, whether coming from local pre-consumer or post-consumer side, forms the upcycling process of creating new products, whereby according to Chen et al. (2015) the design process as such can be seen as "an integral part of sustainable development for its potential to make significant contributions to society: leading social change, realising social values and influencing customer experience". Armstrong et al. (2014: 15) combine design and the social aspect stating that merging both through participatory approaches enables the generation and realisation of new ways allowing change on collective and social levels in contrast of mostly following commercial aims. Difficulties in some cases arose with the production process foremost where used raw materials had a direct negative impact on the standard of the product quality (Project A). Creating a fully circular product or packaging was in some cases hindered by a lack of material accessibility in contrast to conventional materials available but was envisioned as important for future business development. Products are evaluated as part of the brand identity and stand symbolically for social justice and repurpose alike. Confirming Hancock (2009: 8), the narrative behind products builds a strong case for marketing strategies and communication and creates a positive brand association for all projects. Products materialise "strong storytelling that puts together environmental aspect (waste) and social values" (Project C) forming part of the brand's identity to be circular through representing people, planet, and profit.

Independently of individual project structures, participants come from a variety of different cultural backgrounds, whereas in some cases nationalities are not clear. Projects bring people together from countries such as Turkey, India, Nigeria, Ghana, Afghanistan, and Russia as well as project leads/organisers from Italy, Germany, Netherlands, Belgium, Morocco, and Denmark. People in different life stages (students, parents, younger and older adults) work together, in one case allowing young men between 18–22 years old with autism to participate in the production phase located at their social organisation. The combination of fostering social inclusion through supporting social change in participants' lives whilst creating marketable products aligns with Markussen (2017: 162) who identifies both aspects as "tightly coupled" when creating social capital.

Considering the investigated business structures above, the following four assumptions are made as mutual factors shaping all projects' business development:

- Through repurposing waste material whilst creating an inclusive working situation, business structures highlight the two most important aspects for all five business ideas: one, using waste, aligned with the principles of a circular economy whereas the second, inclusivity and multiculturalism, is yet to become a solid part of what can be an expected approach for other SMEs as a more holistic implementation of a social circular economy.
- Creating suitable business structures addressing social change is the result of personal interest and motivation. Belief and vision seem to be the main cause for founding projects matching circular economy principles with social impact.
- Offering circular designed products is assumed to result from a strong interest in tackling social injustice and disadvantage. Products classified as sustainable only would not be "enough" concerning the change project leaders want to achieve with their businesses as they participate in the economy as problem solvers.
- The story around the products created builds on social change driven by business impact and connects the individual with the broader societal context in

which social injustice can be challenged through design and products become the identity of value and vision.

Benefits and challenges

Qualtrix survey data allowed us to compare and identify similarities and differences related to the benefits and limitations of the specific project structures through measurable outcomes. As Figure 10.1 shows, although there are nuances in individual cases where one or the other parameter was considered more or less impactful, a similar emphasis on some challenges becomes visible. This includes the "general benefit for participants" or the "project benefit from cultural diversity," for example, which are ranked high (4–5) by each project or the lack of consumer interest posing a low challenge to their business success (ranked 0–1).

Cultural diversity adds different perspectives to the projects, enriching both sides of the project lead and participants with input and ideas on a creative level. Moreover, coming from multicultural backgrounds, participants benefit from learning about other life concepts offering insights into similarities and differences which directly impact the "openness towards other people" (Project B). In addition to highlighting social justice, the multicultural experience gives purpose to participants but also on an extended level to customers and "brings the heart to the project" (Project D). Overall, benefits are foremost seen on an individual,

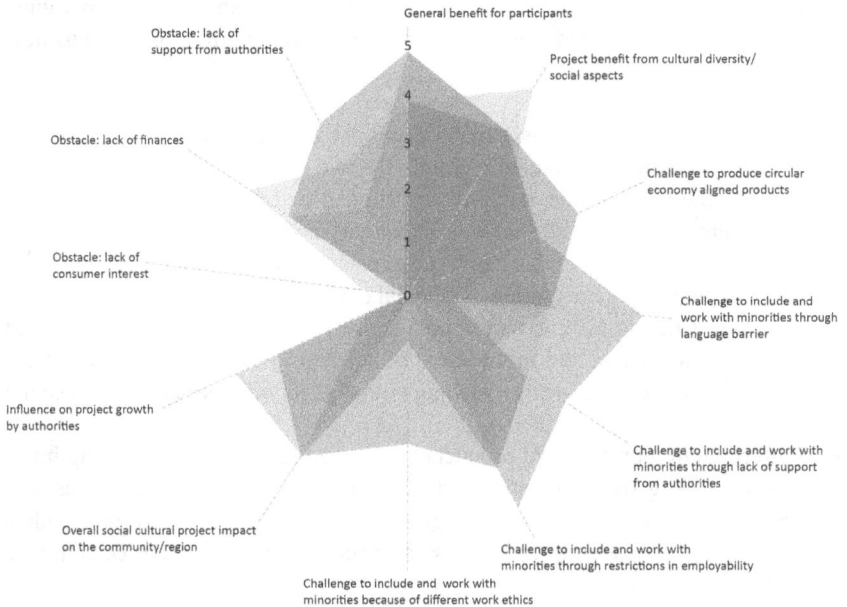

Figure 10.1 Qualtrix survey outcomes. Each shade corresponds to the data of one project.

personal level responding to the founder's visions and beliefs when projects were developed at the beginning.

One of the mutual challenges rooting in working with multicultural participants is the language barrier as well as shyness at the beginning of the project – especially where due to other cultural structures men and women are not used to working together. To overcome the latter in regard to creating a positive workshop setting, one of the project leads turned on music and successfully used dancing and hands-on creative work as an icebreaker (Project A). Another challenge mentioned across projects is different work ethics or the different handling of rules which result foremost from participants' different cultural habits. In some cases, these cause a different understanding of qualitative reasonable product outcomes, a lack of discipline, and furthermore a lack of communication in situations that could be seen as normal in other working environments when calling in absent for example. Emphasising lowering hierarchies, project leads actively show how making mistakes or struggling to complete a task is meant to be okay as participants are afraid of rejection or have low self-esteem due to difficult life circumstances. As the legal situation doesn't allow some participants to be paid for their work, project leads show their value through mutual spare time activities to "give something back" (Project D).

Common difficulties on a business level can be categorised into legal challenges raising from tight rules, material sourcing, expenditure of time, and promoting the mindset behind projects to potential customers. Legal challenges like certificate achievement or employment difficulties lead to a deceleration of production and sales. Generally, processes are at times more time consuming due to the special needs of some participants who require a different understanding and work pace. Or processes are less stable where leaving participants caused a discontinuing working environment. As the projects are dependent on external suppliers of materials, scarcity, especially in times of coronavirus, led to a shortage of material which however can be overcome with having multiple suppliers. Not being able to look at other existing businesses as a form of role model due to scarcity of similar approaches is mentioned by Project B as a difficulty for the project set up. Finally, all products materialise a mindset of inclusion, fairness, and sustainability which can be challenging to promote, especially amongst elderly clients, but are likewise envisioned as a positive means of "better things come from challenges" (Project E).

Supportive measures facilitating project growth differ from case to case. Overall, they can be divided into four main areas of 1) funding, 2) social support, 3) legal advice, and 4) promotion through and within local communities. Receiving governmental funding or having the possibility to apply for a grant as a business foundation is seen as supportive across all projects. Another required support relates to the social level. Participants do not only face challenges of being inexperienced with work life as such but are, furthermore, exposed to difficulties resulting from dealing with bureaucracy, for example. Having access to information about where to get help for specific problems from the project lead side is said to foster social inclusion (Project B). Depending on the products developed, receiving legal advice where certifications are mandatory to be able to offer products on the market, for example, enables sales and project development. Cultivating connections to local communities aiming to promote projects is also evaluated as supportive for facilitation.

Resulting from the first part of the survey the following four assumptions are identified as mutual project benefits and limitations:

- Similar to the processes of design thinking where teams are put together explicitly from different backgrounds aiming at contributing with different perspectives, the multicultural experience can be seen as an approach to creating unique teams where business value increases the more diverse viewpoints are brought together.
- Cases show that multicultural and inclusive working environments face day-to-day challenges on different levels ranging from language barriers, legally not being able to pay workers, and different work ethics to guiding inexperienced employees in how to work beyond the actual tasks.
- Projects are challenged with issues such as the need for certificates, material sourcing, expenditure of time due to unique needs, and promotion of the materialised mindset, especially amongst elderly customers. Also, the dependency on other stakeholders in terms of material supply sometimes inhibits the production process.
- Similar to other entrepreneurial businesses, projects focusing on social change face challenges such as a lack of funding, legal advice requirements, and a lack of promotion on a local level. These may differ from legal business type to business type. In addition, supportive measures related to the inclusive and multicultural setting are required such as putting together a general how-to guide addressing daily life tasks ranging from understanding bureaucratic handlings to supporting time management.

Social impact of multicultural and inclusive communities

Adding to the earlier discussed definition of positive social change Stephan et al. (2016), furthermore, specify the activity as "proactively initiated through [...] market-based organisations [...] operating in a competitive environment" whilst the act of achieving positive social change aims at involving the transformation of individuals and groups. Activism by project leads intending to drive change for participants results from personal belief or vision as evidenced in Chapter 1. However, across all five projects, all people involved experience how their work unites them and impact becomes evident for everyone engaged as a means of integration. People from disadvantaged backgrounds have a feeling of belonging and a daily purpose which continues to have a positive ripple-down effect on their families, children, and social environments (Project B). Projects help participants to integrate themselves, learn new skills, and gain support in a learning process beyond the actual work tasks identified by Julier and Kimbell (2019) as part of the real change which "happens in the slow, tricky, and political work of implementation". Whilst project leaders benefit from new perspective-taking and approach enriching their everyday life, local communities are included by personal interactions in markets or in local shops, initiating learning more about ongoing projects in their hometown and creating a further interconnected cultural exchange.

Multicultural experience is positively related to performance in solving a problem that requires insight and to producing creative ideas without being confined to the widely known. It also predicts creativity-supporting processes such as the tendency to access unconventional knowledge from memory and to recruit ideas from foreign cultures for creative idea expansion (Leung et al. 2008). Moreover, the relationship between multicultural experience and creativity is particularly strong when people adapt and are open to these new experiences and when the creative context deemphasises the need for firm answers or morality concerns. Benefitting from cultural diversity seems to work both ways for participants and project leads alike. Learning processes happen on different levels with no exemption for project leads as participants are not taught in a classical sense but everyone learns new skills together, prejudices are reduced, and participants open up to people involved. Whereas in one case being set up through the student NGO Enactus, the project is built with students from different studies and ethnic backgrounds. Therefore, it was diverse in its framework already and brought a shift in perspective from the beginning. Furthermore, craftwork as such is argued to be healing and helping participants (Project C). Another important aspect is the attention being given to people making them feel part of this world. On a broader level, attention is also drawn to further institutions involved such as social organisations for example (Project D).

The following three key points identify the projects' impact on a social level and highlight the contribution coming from the specific multicultural background:

- Projects give participants purpose, refine their dignity, increase their self-confidence, and connect them with others fostering integration and immigration. Cultural diversity opens up mindsets and creates acceptance. A sense of belonging evolves through mutual learning and participation as such. Attention towards the individual increases but furthermore towards the structural settings of the concepts overall highlighting social disadvantages or social injustice.
- All five projects unite participants with project leaders and the local communities while everyone benefits from this interaction in different aspects. While the integration process for people from disadvantaged backgrounds is facilitated, triggering a generation effect, project leaders gain new perspectives, enriching their work and private life. Meanwhile, consumers have valuable encounters in places like markets or in small shops, learning more about project developments in their local area.
- Bringing employees and students from different cultural backgrounds into the same team or department provides one form of multicultural experience

that can potentially make people more facile at creative problem solving and idea generation.

Shifting personal interest to mainstream activism

Drawing from the previous research it is argued that transformation processes such as the transition towards a circular economy are always rooted in changing individual perceptions and practices. Thus, the focus lies on how business and community actors through their behaviour become part (or not) of new collective practices, which are changing routines in economic processes. The implication is that with changing to circular business models, the role of individuals changes from ownership to use (Ellen MacArthur Foundation 2013) indicating a passive role. Individuals, however, are key actors in the change towards a social circular economy concerning consumer responsibility (Ghisellini et al. 2016), changing consumption practices (Hobson 2016), changes in norms and practices relating to ownership, and even in influencing policy through voting behaviour, to name a few. Given the limited knowledge about the willingness of citizens to change their consumption practices, circular economy research should deal with how values and norms influence the change process, or how different policy measures or business models can influence their behaviour (Hobson 2016).

Embedded into the principles of a circular economy, value creation as of today mainly takes place on the product level when businesses concentrate on designing out waste or keeping products in the loop. Hence, many businesses do not carefully consider their opportunities to drive social change. Although design is acknowledged as a powerful tool to create better futures, the way design is actually practised often takes the social backseat. Project development uncertainty or the prospect of economic growth delay coming with business parameters similar to the cases discussed is considered causing a lower percentage of brands aiming for social change. However, it needs to be added that business development is often driven by unique interests which do not naturally include social aspects but can be influenced by others such as monetary values, for example.

As a basis for the following proposal aiming at supporting a shift from individual interest into mainstream activism, the average of the survey outcomes (Figure 10.2) below was used to allow more generalised and industry transferrable suggestions:

Average data show the positive impact on participants and the project is equally high, similar to the overall social-cultural impact on the community or region. Likewise, challenges are evaluated as being moderately high on some levels (producing circular economy aligned products, language barrier, lack of support from authorities, and restrictions in employability). Taking obstacles into account such as a medium-high lack of finances and lack of support from authorities, one might state project leaders have a very strong individual interest and/ or motivation to develop and sustain their social business ideas. This aligns with Zahra et al. (2009) who characterise social entrepreneurs as innovative individuals. Considering the above it must be asked how conditions for SMEs can be set

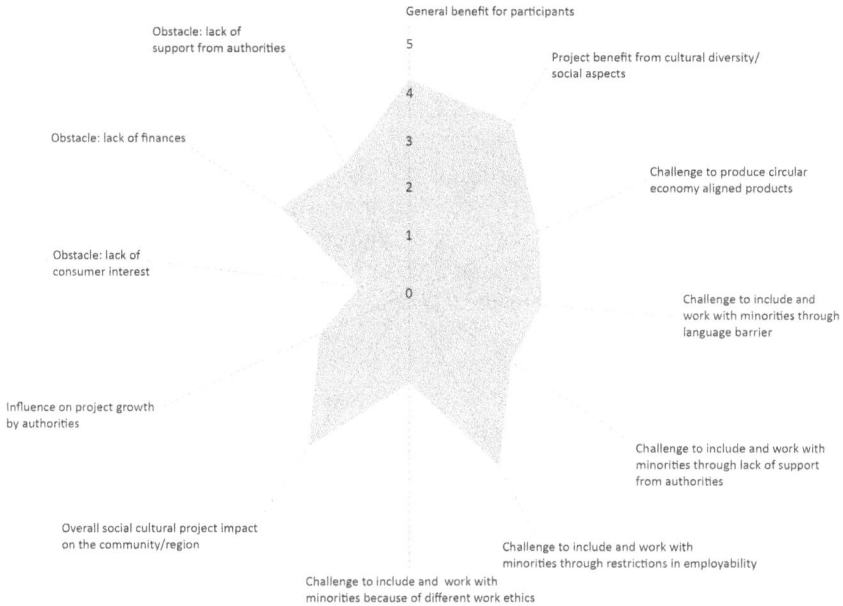

General benefit for participants

Obstacle: lack of
support from authorities

5

Project benefit from cultural diversity/
social aspects

4

Obstacle: lack of finances

3

Challenge to produce circular
economy aligned products

2

1

Obstacle: lack of
consumer interest

0

Challenge to include and
work with minorities through
language barrier

Influence on project growth
by authorities

Challenge to include and work with
minorities through lack of support
from authorities

Overall social cultural project impact
on the community/region

Challenge to include and work with
minorities through restrictions in employability

Challenge to include and work with
minorities because of different work ethics

Figure 10.2 Average Qualtrix survey outcome of all five projects.

in such a way that the individual factor of motivation can be partially levelled out through clear additional support on a regulatory, political level if there is little or no belief or vision to create social business structures?

Resulting from the previous investigations it is proposed that activities to increase the social impact of SMEs should be shared across various responsibilities from individual commitment to political, communal level. Could there be such a thing as an *Entrepreneurial Social Circular Business Toolbox* jointly offered by governments and communities aiming to ease the founding and development process through an already existing basic business framework? And if so, would governments be interested at all as it might be evaluated as admitting systems failure and shifting responsibility to the individual? Following the challenges revealed, the following paragraphs show potential components of a toolbox guiding businesses aiming at building their framework around multicultural and inclusive pillars.

Similar to other entrepreneurial businesses, circular economy projects focusing on social change face challenges such as funding, legal advice, and lack of promotion on a local level. These may differ from business type to business type. Additionally, where material sourcing and material supply solely depends on other stakeholders it can lead to material scarcity inhibiting the production process. In some cases, due to the unique needs of participants, the scale and growth of production are limited. From a materialised customer perspective, the uniqueness of each product due to repurposing waste material can hinder the appreciation of the product itself. According to one interviewee, the full implementation of circular economy principles requires intense research. Lacking time and capacity to find suitable raw materials or business partners to develop circular products could be tackled by local institutions performing as *Circular Business Angels,* offering support in networking or guidance in circular thinking whilst following the personal business motivation such as economical ambition, for example. Local platforms such as www.circular .berlin can accelerate business exchange towards a more inclusive economy.

Projects benefit in many ways from their multicultural and inclusive settings. Therefore, it is assumed that purposely developed working environments where creative problem solving is based on diverse mindsets shaped through different life concepts, cultural backgrounds, and age groups impacting businesses positively. Participants and brands perform multicultural design thinking structures allowing broader business perspectives and individual growth alike. Supporting inclusive structures from the community side through collaborative exchange with local institutions resulting in *Inclusive Leadership Training,* for example, could increase interest in building diverse structures. Moreover, empowering people to develop inclusive cultures where everyone feels respected and valued promotes an inclusive mindset, exploring new perspectives and practices.

A direct link can be seen between the challenges to include and work with minorities through restrictions in employability and the lack of support from authorities. Being limited to creating business structures comparable to other SME businesses where limitations amongst others might be rooted in lacking the economic power to employ workers, these projects face additional restrictive legal settings. For instance, Project B had problems employing skilled migrants due to lacking or unacknowledged certificates while another project was challenged by not being able to employ refugees due to their asylum status. Thus, policies creating *Flexible Forms of Employability* could ease the process of integrating minorities into team structures on a long-term basis. As a result, where successful inclusive businesses can tackle the lack of finances leading to project growth, SMEs could increase an overall positive impact on communities and local regions.

Challenges resulting from the multicultural background reach beyond the actual day-to-day business. Projects develop creative handling of supporting different mentalities, cultures, and socially disadvantaged participants to create a steady working environment. Especially as none of the project leads is a social worker, it can be assumed that all have a personal motivation to support their participants aiming at social inclusion not limited to the workplace. Support beyond working life includes helping with letters from administrative offices, assisting in making doctor appointments and overall psychological encouragement. Building upon this interest addressing the how-to of life in a foreign environment in form of *Yellow Pages,* project leads could be prepared better and more efficient in support. Also, knowing whom to approach when facing challenges eases initial barriers that projects might face.

As projects give participants purpose, refine their dignity, increase their self-confidence and connect them with others fostering integration and immigration it is argued that cultural diversity opens up mindsets and creates acceptance. A sense of belonging evolves through mutual learning and participation as such. Attention towards the individual increases but furthermore towards the structural settings of the concepts overall highlighting social disadvantages or social injustice. Participants are united with local communities and project leaders and experience beneficial interaction in different aspects. Therefore, instigating *Community Events* connecting local people and participants can increase the overall awareness of injustice and social disadvantages. Talks, discussions, or activities like clothing repairs can promote projects and foster integration and knowledge exchange within local communities alike.

In summary, the *Entrepreneurial Social Circular Business Toolbox* comprises different aspects to support social business development processes with application possibilities in other industry branches:

- Circular Business Angels
- Inclusive Leadership Training
- Flexible Forms of Employability
- Yellow Pages
- Community Events

The adaptation of the basic framework can be seen as a pick-and-choose offer accommodating individual brand needs.

Conclusion

The growing impact of environmental exploitation and climate change on global migration movements adds to the existing demand for societal restructuring on the local level caused by a growing income gap or lack of employment opportunities, for example. Amongst other ways, sustainable entrepreneurship is seen as a potential solution to simultaneously balance "economic health, social equity and environmental resilience" (Cohen and Winn 2007: 34) aiming at overcoming the dichotomy between present and future generations. With 98% of all European businesses being SMEs (European Commission 2020), there seems to be a huge potential to drive required social change whilst tackling environmental threats when business structures recognise the circular economy as a powerful tool to symbiotically function as a social change facilitator through design. All five European projects investigated in this case study show how creativity and craftsmanship as a proactive, underpinned process of their businesses can perform as a "vehicle for empowerment" (Sheperd and Patzelt 2011) fostering positive social change as problem solvers. As a second mutual policy aligned with the principles of a circular economy, all utilise pre- and post-consumer waste for product development to enhance the projects' impact on an environmental level. Interview and survey outcomes have highlighted similarities and differences in the challenges and benefits projects face due to their multicultural and diverse settings. These were identified and evaluated to define a framework proposal as a supportive tool for other companies to accelerate social equity and justice alongside the interest in economic activities addressed through a circular economy-based structure.

The *Entrepreneurial Social Circular Business Toolbox* framework suggests five components aiming at lowering the barriers for the development of inclusive, multicultural economies in particular. Drawing from the explored challenges which can be divided into the areas of funding, social support, legal advice, and promotion through and in local communities, the Entrepreneurial Social Circular Business Toolbox offers support that relies on governments and communities alike to take over responsibility and play their part in creating future-directed local structures. To summarise, these are: first, offering supportive measures to create effective circular business strategies and local networks through *Circular Business Angels* aiming at developing fully circular products. Second, *Inclusive Leadership Training* can accelerate business development through supporting mindsets towards building an inclusive economy. Third, governments are required to create *Flexible Forms of Employability* to ease the process of working with inclusive and multicultural team structures on a business-related basis. Fourth, as project leads might face tasks beyond the actual day-to-day business *Yellow Pages* could offer a how-to guide developed by municipalities. And finally, fifth, initiating *Community Events* to offer interaction and exchange with residents and local businesses. To define the toolbox in greater detail, further research should focus on elaborating detailed planning of the individual components as a collaborative approach between the stakeholders mentioned above.

Research results show that the reasons for creating an inclusive and multicultural business structure are based on personal vision and belief in social innovation. Project benefits are also widely seen on a personal level where cultural diversity adds different perspectives, mutual learning, and purpose enriching the lives of everyone involved which, in reverse, might outweigh the challenges arising. Therefore, supportive measures can potentially only be fruitful where people already consider the need for social change as part of their project idea or as a means of personal growth, emphasising the role of individuals. As a consequence, it is argued that the clearer the roads to take the lower the barriers to creating inclusive and multicultural business structures will be. Adding clarity also includes initiating a wider political discourse about the feasibility of such measures in reliance on political engagement. These require responsibility acknowledgement by governments and municipalities in the first place.

In conclusion, it is argued that although the initiation of social enterprises depends on the personal motivation and activism of project founders, additional supportive measures offered by governments or local communities could ease the development and maintenance of social businesses for less socially interested entrepreneurs. By means of the suggested toolbox, governments could seize supportive measures thereby facilitating policy making towards a more inclusive working environment. Furthermore, multicultural and diverse structures need to be valued by municipalities as a powerful way to positively shape societies through a ripple-down effect by design. Business setup can function as a practice-based design thinking tool offering perspectives from different (cultural) backgrounds tackling ecological and social problems. By addressing global problems through a local solution, systems can be created of which, according to the circular economy definition, wider societies benefit, considering the underlying term "by design" is further specified as "by-design activism".

References

Apostolopoulos, Nikolaos, Haya Al-Dajani, Diane Holt, Paul Jones, and Robert Newbery, eds. 2018. *Entrepreneurship and the Sustainable Development Goals*. 1st ed. Contemporary Issues in Entrepreneurship Research, Vol. 8. Bingley: Emerald Publishing Limited, pp. 1–7.

Armstrong, Leah, Jocelyn Bailey, Guy Julier, and Lucy Kimbell. 2014. *Social Design Futures: HEI Research and the AHRC*. Brighton: University of Brighton. Accessed 01 November 2020. https://cris.brighton.ac.uk/ws/portalfiles/portal/341933/Social-Design -Report.pdf.

Chen, Dung-Sheng, Lu-Lin Cheng, Caroline Hummels, and Ilpo Koskinen. 2015. "Social design: An introduction." *International Journal of Design* 10 (1): 1–5. http://www .ijdesign.org/index.php/IJDesign/article/view/2622/723.

Choi, Youngok, Jea Hoo Na, Andrew Walters, Busayawan Lam, John Boult, P. W. Jordan, and Stephen Green. 2018. "Design for social value: Using design to improve the impact of CSR." *Journal of Design Research* 16 (2): 158. doi:10.1504/ JDR.2018.10014196.

Cohen, Boyd, and Monica Winn. 2007. "Market imperfections, opportunity and sustainable entrepreneurship." *Journal of Business Venturing* 22(1): 29–49. doi:10.1016/j. jbusvent.2004.12.001.

Daly, Herman E. 2007. *Ecological Economics and Sustainable Development, Selected Essays of Herman Daly*. Cheltenham: Edward Elgar Publishing Limited.

Ellen MacArthur Foundation. n.d. "Concept." Ellen MacArthur Foundation. Accessed 01 November 2020. https://www.ellenmacarthurfoundation.org/circular-economy/ concept.

Ellen MacArthur Foundation. 2013. "Towards a circular economy: Economy and business rationale for an accelerated transition." Accessed 16 November 2020. https://www .ellenmacarthurfoundation.org/assets/downloads/publications/Ellen-MacArthur -Foundation-Towards-the-Circular-Economy-vol.1.pdf.

Ellen MacArthur Foundation. 2015. "Towards a circular economy: Business rationale for an accelerated transition." Accessed 17 November 2020 https://www.ellenmacarthurf oundation.org/assets/downloads/publications/TCE_Ellen-MacArthur-Foundation_26 -Nov-2015.pdf.

European Commission. 2020. "Unleashing the full potential of European SMEs." European Commission, 10 March 2020. https://ec.europa.eu/commission/presscorner/detail/en/fs _20_426.

Ghisellini, Patrizia, Catia Cialani, and Sergio Ulgiati. 2016. "A review on circular economy: The expected transition to a balanced interplay of environmental and economic systems." *Journal of Cleaner Production* 114: 11–32. doi:10.1016/j.jclepro.2015.09.007.

Hancock, Joseph II. 2009. *Brand Story: Ralph, Vera, Billy and Other Adventures in Fashion Branding*. New York: Fairchild Books.

Harding, Thomas S., and Dean Whitehead. 2015. "Analysing data in qualitative research." In *Nursing & Midwifery Research: Methods and Appraisal for Evidence-Based*

Practice, edited by Zevia Schneider and Dean Whitehead, 127–142. 4th ed. Mosby: Elsevier.

Hazarika, Akanshiya. 2013. "Corporate social responsibility and workplace democracy: Emerging issues and perspectives." *Journal of Management & Public Policy* 5 (1): 27–40. http://jmpp.in/wp-content/uploads/2016/01/Anjana-Hazarika.pdf

Hobson, Kersty. 2016. "Closing the loop or squaring the circle? Locating generative spaces for the circular economy." *Progress in Human Geography* 40 (1): 40–104. doi:10.1177/0309132514566342.

Holt, Diane, and David Littlewood. 2018. "Social entrepreneurship in South Africa: Exploring the influence of environment." *Business and Society* 57 (3): 525–561. doi:10.1177/0007650315613293.

Julier, Guy, and Lucy Kimbell. 2019. "Keeping the system going: Social design and the reproduction of inequalities in neoliberal times" *Design Issues* 35 (4): 12–22. doi:10.1162/desi_a_00560.

Klepp, Silja. 2017. "Climate change and migration." In *Oxford Research Encyclopedia of Climate Science*. New York: Oxford University Press. doi:10.1093/acrefore/9780190228620.013.42.

Leung, Angela K., William W. Maddux, Adam D. Galinsky, and Chi-Yue Chiu. 2008. "Multicultural experience enhances creativity: The when and how." *The American Psychologist* 63 (3): 169–81. doi:10.1037/0003-066X.63.3.169.

Margolin, Victor. 2015. "Social design: From Utopia to good society." In *The Social Design Reader*, edited by Elizabeth Resnick, 17–30. London: Bloomsbury.

Markussen, Thomas. 2017. "Disentangling 'the social' in social design's engagement with the public realm." *International Journal of CoCreation in Design and the Arts* 13 (3): 160–174. doi:10.1080/15710882.2017.1355001.

Mebratu, Desta. 1998. "Sustainability and sustainable development: Historical and conceptual review." *Environmental Impact Assessment Review* 18: 493–520. doi:10.1016/S0195-9255(98)00019-5.

Moulert, Frank, Diana MacCallum, and Jean Hillier. 2013. Social innovation: Intuition, precept, concept, theory and practice. In *The International Handbook on Innovation*, edited by Moulert, Frank, MacCallum, Diana, Mehmood, Abid, and Hamdouch, Abdelillah, 13–24. doi:10.4337/9781849809993.00011.

Murray, Alan, Keith Skene, and Kathryn Haynes. 2017. "The circular economy: An interdisciplinary exploration of the concept and application in a global context." *Journal of Business Ethics* 140: 369–380. doi:10.1007/s10551-015-2693-2.

Pedersen, Esben, Rahbek Gjerdrum, Wencke Gwozdz, and Kerli Kant Hvass. 2018. "Exploring the relationship between business model innovation, corporate sustainability, and organisational values within the fashion industry." *Journal of Business Ethics* 149 (2): 267–84. doi:10.1007/s10551-016-3044-7.

Redclift, Michael. 2005. "Sustainable development (1987–2005): An oxymoron comes of age." *Sustainable Development* 13: 212–227. doi:0.1002/sd.281.

Stephan, Ute, Malcolm Patterson, Ciara Kelly, and Johanna Mair. 2016. "Organizations driving positive social change: A review and an integrative framework of change processes." *Journal of Management* 42 (5): 1250–1281. doi:10.1177/0149206316633268.

Shepherd, Dean A., and Holger Patzelt. 2011. "The new field of sustainable entrepreneurship: Studying entrepreneurial action linking 'What Is to Be Sustained' With 'What Is to Be Developed'." *Entrepreneurship: Theory and Practice* 35 (1): 137–163. doi:10.1111/j.1540-6520.2010.00426.x.

World Commission on Environment and Development. 1987. *Our Common Future*. Oxford: Oxford University Press. https://sustainabledevelopment.un.org/content/documents/5987our-common-future.pdf

Zahra, Shaker A., Erik Gedajlovic, Donald O.Neubaum , and Joel M.Shulman . 2009. "A typology of social entrepreneurs: Motives, search processes and ethical challenges." *Journal of Business Venturing* 24 (5): 519–532. doi:10.1016/j.jbusvent.2008.04.007.

11 Contemporary art and cosmovisions of Brazilian indigenous peoples

Potential influence on CE and SSE practices?

Luciana Benetti Marques Valio

In this text, I intend to analyse the contemporary artwork *Dois pesos, duas medidas* (*Double Standard*) by Lais Myrrha, which instigates us to reflect on the distinct ways of life and projects of societies that understand nature and the environment differently. On the one hand, human societies pursue in an exploratory manner for the benefit of their own comfort (and progress), and on the other, in an integrated manner, establish a harmonious and sustainable relationship with nature (and life on the planet). In other words, human life and coexistence with the natural environment is not merely a matter of addressing the processes in the productive chain of construction methods, but primarily one of considering the underlying worldview beneath those processes. Therefore, via interpreting Myrrha's work of art, I suggest that changes that must be made in the world to mitigate the effects of climate change will only really take place if we shift the epistemological bases. Thus, if we really want to make profound transformations in the world, which are becoming increasingly urgent, we need to review our worldview.

Aníbal Quijano suggests that the epistemological basis of the Western Modernity is established via colonisation, thus it is important to highlight that the colonisation of the Americas aided the solidification of the capitalist system, which in return developed by the exploitation and exploration of natural resources in Latin America and in other continents. The impact of colonisation has underlined the differences between indigenous peoples' ways of life and the new order imposed by Western society. I will use the term epistemicide, coined by Boaventura de Sousa Santos, in the Brazilian case to elucidate the destruction of the cultural and natural life of indigenous peoples in Brazil. In other words, the relationship between nature and indigenous peoples was destroyed, with the aim of a civilising process, which subjugated these cultures and even the peoples themselves.

In this vein, Laís Myrrha's artwork for the Thirty-Second São Paulo Art Biennial, in 2016, may contribute significantly to our understanding of the social and cultural aspects of circular economy, via the demonstration of the difference between the indigenous and Western worldviews. In this sense, Myrrha's work of contemporary art allows readings to go beyond its materiality, and can be contextualised in relation to the elements that are given, such as, for example, the

DOI: 10.4324/9781003255246-11

title, the materials used, the form/format, where it was done and/or is installed, the event of which it is a part, the political context, and the curatorial argument promoted. Our interaction with it, which entails investigating and questioning aspects from the smallest details in the work to the wider context, induces us to better understand the world around us.

In addition to this, it is important to highlight the curatorial proposal for this edition of the São Paulo Art Biennial, which emphasised the ecological crisis and pointed out the indigenous people's issues. Moreover, the political context, which we are still witnessing and suffering its aggravating effects, is also contributing to the dismantling of the welfare state and the democratic institutions. All this context was presented in the curatorial proposal of the show. And Myhrra's work occupies the exhibition main space inside the Biennial Pavilion, which is known as "o vão" (the span), because the artist's work could be seen from all three floors of the Pavilion.

In this regard, contemporary art could let us comprehend aspects that are not explicit. I chose Myrrha's artwork *Double Standard* with the intention of highlighting a conflictual relationship that has been developed in Brazilian society since the sixteenth century, when the colonisation of the Brazilian territory began. By comparing the building methods used in industrial civil construction and for building the indigenous dwellings of the Upper Xingu peoples, the artist sheds light on how different worldviews affect the environment in different ways.

She compares the two constructive processes, and this allows us to analyse the different impacts on the environment of those models – especially in the civil construction industry, where the take-make-use-dispose-pollute system is extremely harmful to nature. In contrast, the natural integrated system used by indigenous peoples can be considered an example of sustainability. This system could even be a reference for the circular economy practices, as well as for the social and solidarity economy, since the social relations of the indigenous peoples are based on communal and collaborative principles in all processes of daily life, emphasising features of a non-hierarchical society (Figure 11.1).[1]

I will begin this explanation by looking at the title of Laís Myrrha's work. This piece was commissioned especially for the Thirty-Second São Paulo Art Biennial in 2016. The title in English, *Double Standard*, already declares the content of injustice, referring to the different treatment of people who should be treated the same. In Portuguese, the artist used the popular saying "Dois pesos, duas medidas" to denote the context of injustice, which in English translates into the archaic expression "two weights and two measures". This expression has biblical origins, specifically in Deuteronomy 25:13–16, which reads:

You shall not have in your bag two kinds of weights, a large and a small. You shall not have in your house two kinds of measures, a large and a small. A full and fair weight you shall have, a full and fair measure you shall have, that your days may be long in the land that the Lord your God is giving you. For all who do such things, all who act dishonestly, are an abomination to the Lord your God.[2]

Figure 11.1 Dois pesos, duas medidas (*Double Standard*), Laís Myrrha. [View from above.] Concrete, brick, mortar, tile, glass, pipes, conduits, wires, metal, wood, piassava, and compacted earth. 800 x 300 x 300 cm. Commissioned by the São Paulo Biennial Foundation for the Thirty-Second Bienal (2016). Image credit: Luciana Valio.

This biblical passage refers to trading, in which a moral value is expressed: honesty. However, more than the quality of being honest (a quality relative to an individual), the passage pertains to social relations, where different powers are established, especially when it says "For all who do such things…are an abomination"; in other words, injustice is committed by those who have the power to do it and, therefore, to be dishonest. In this regard, it becomes a political question whereby individuals with different powers, upon engaging in a trade, act unfairly.

With this in mind, the title of the artwork offers an indicator of what it is about: injustice. During the period when the work was produced, in 2016, President Dilma Rousseff of Brazil was impeached and removed from office. Although *Double Standard* (2016) may allude to the political injustices committed in the country, the work specifically refers to the injustices committed against the indigenous peoples in Brazil since colonisation: genocide, enslavement, devaluation of their cultures, erasure of their history, misappropriation of their lands, etc.

Before describing Laís Myrrha's work in more detail, and in order to contextualise the stance intended in this chapter, it is important to highlight colonisation as part of coloniality, a concept introduced by Aníbal Quijano. In the 1980s and 1990s, Quijano, a Peruvian sociologist, coined the term "coloniality" to designate the "colonial matrix of power". This refers to the order that has been established in Western society since the Renaissance, in the wake of the European invasions of the Americas and the enslavement of Africans to work in the Americas. For Mignolo (2017), modernity is sustained by the existence of coloniality.

Quijano (2014) underlines that one of the fundamental elements of coloniality/modernity/Eurocentrism was the separation from "reason" and from "nature". Therefore, at the beginnings of modern science, human beings (Europeans) armed with reason (and bearers of the truth/knowledge) are differentiated from nature, which is wild and, consequently, subject to being civilised and exploited. Hence, the idea of nature and culture being opposites, founded on the Renaissance by Francis Bacon, in his *Novum Organum* (1620),

> in which he proposed a reorganization of knowledge and clearly declared that *nature* was *there* to be tamed by man. During this pre-Industrial Revolution period, Western Christians asserted their control over knowledge about nature, disqualifying all other existing and equally valid concepts, and ignoring concepts that contradicted their own understanding of nature. At the same time, they engaged in an economy of extracting raw resources (gold, silver and other metals) for a new type of global market.
>
> (Mignolo 2017: 7)

With the invasion of the New World, therefore, a new economy emerged, which consisted of reinvesting profits to increase production, based on the continuous reproduction of the resources. The invention of nature separated from the human being thus enabled and, above all, justified its exploitation due to the need for natural resources, later intensified in the productivist ethic of the Industrial Revolution.

Furthermore, Quijano (2014) emphasises how tightly bound the idea of race was to being a part of nature, as a pretext for the creation and exploitation of "inferior races". Such a distinction between races became necessary for the maintenance of this new economic system. Previously, work had been performed through the need to maintain life. However, to sustain the new system, work became an activity to generate the accumulation of wealth. Especially since the slave trade was a very lucrative activity for bolstering the Industrial Revolution.

Whereas Bacon instigated the understanding of nature as separate from the human being, it was through the Industrial Revolution that nature became a supplier of raw materials and was converted into "natural resources", intensely exploited for the growth and development of the capitalist system. Therefore, the idea of progress and modernisation – tied to the vision of development as improvement to life, or rather, as the only possible way of life, that should be desired and sought by all, especially those who were subjugated by modernity – leads to epistemicide, a term coined by Boaventura de Sousas Santos. This equates to the

> suppression of local knowledge perpetrated by an alien knowledge [...] Indeed, under the pretext of the "colonizing mission", the colonization project sought to homogenize the world, obliterating cultural differences.[3]. This led to the loss of vast social experience and reduced the epistemological,

cultural and political diversity in the world. To the extent that they survived, those experiences and that diversity were subjected to the dominant episte-mological norm: they were defined (and often ended up being self-defined) as local and contextual knowledges, which could only be used in two cir-cumstances: as raw material for the advancement of scientific knowledge; and as instruments of indirect government, instilling in the peoples and their practices the credible illusion of being self-governed. The loss of a genuine self-reference was not only a gnoseological loss, but also, and primarily, an ontological loss: inferior knowledge belonging to inferior beings.

(Santos 2010: 16–17).

That given, it demonstrates that these knowledge were not considered legitimate and true, stamped by modern science; they were deemed as superstition, and scorned, together with the cultures and peoples who held them. The indigenous wisdom was subjugated and Western/modern epistemology was transformed into merchandise, as a model, in its economic system, to be exported and acquired by non-Western peoples.

The purpose of presenting this perspective is derived from the fact that in *Double Standard* one can glimpse how the indigenous peoples and their knowl-edge are devalued. Laís Myrrha, through the title of her artwork, was referring to the countless injustices that were, and still are, inflicted on the indigenous com-munities. However, this allusion to the indigenous peoples is manifested in the materiality of the work. The work consists of two towers placed in the main con-course of the São Paulo Biennial Foundation Pavilion (Figure 11.2). One tower refers to the building methods of the indigenous communities of the Upper Xingu (Mato Grosso state, Brazil), and the other tower is ascribed to modern civil con-struction found in Brazilian urban centres.

When walking up the ramp to the ground floor of the pavilion, the base of the first tower can be seen from afar. Gradually, as we walk toward the main con-course, the tower emerges into view and seems to grow together with the space. And, in the same way, the second tower also appears. When we reach the main concourse, both towers are in full view, revealing the vastness of the pavilion's internal space; although very tall, the towers fail to reach the second floor of the building.

The first tower seen upon entering the exhibition space was built with rammed earth, treated blue gum posts, bamboo, and *sapé* thatch. Just beyond this one is the second tower, which is composed of concrete, bricks, mortar, pipes, tiles, and glass. The visual impact caused by the two towers installed in the main concourse of the pavilion leads to an inevitable comparison, possibly due to their dimen-sions. For it establishes a relationship of scale between the architectural space of the pavilion and the towers that occupy a highly visible area that stands out from the different floors.

The act of stacking the materials used in the respective building methods was the formal synthesis that the artist chose to represent housing of the Upper Xingu and of Brazilian civil construction. It should be considered that "an important

Figure 11.2 Dois pesos, duas medidas (*Double Standard*), Laís Myrrha. [Front view.] Concrete, brick, mortar, tile, glass, pipes, conduits, wires, metal, wood, piassava, and compacted earth. 800 x 300 x 300 cm. Commissioned by the São Paulo Biennial Foundation for the Thirty-Second Bienal (2016). Image credit: Luciana Valio.

aspect of her [the artist's] creative process is the selection and precise use of materials, which reveal her attention to their ability to signify, to function symbolically and condense narratives" (Volz and Rebouças 2016: 104). It would seem that with this approach, the artist chose to create a monument, which due to its format, can be considered a totem. So, it is not an architectural space to be entered, on the contrary, the work should be viewed from outside and be seen from afar. In formal terms, there are differences between construction materials and methods to which she refers. She used the quantity of materials required to build an indigenous dwelling of the Upper Xingu and also enough materials to build a single-family detached house. And she stacked the materials for each building method into two towers roughly eight metres tall. The artist opted to use only the structural materials of the houses, disregarding finishing materials. Hence, there is no social distinction highlighted by the type of paint, tiles, porcelain, metals, or electrical installations, in the economic sense.

The distinction made in the work refers to the indigenous question, which can be read on several levels, from the political approach to indigenous issues adopted by our governors, to understanding how the indigenous building method follows a logic that is contrary to that of the civil construction industry. The artwork incites various reflections, such as, what value ("weight") is given to indigenous issues? And, what impact ("weight") does the civil construction industry have on the environment?

These are different construction logics, used for different purposes. The indig-enous housing, which the Kamayurá call "ok", or *oca*, is not only a house, in the narrow sense of a residence; the *oca* is a social space, of relations and chores. On the other hand, the modern house of our society is the unit for living, often only used as a dormitory, as our social and productive lives take place elsewhere, often in the urban centres far from our homes. Hence, Myrrha's work addresses "two building methods that embody ways of life and two different projects of society that, even if they are possibilities for construction, already declare their forms of ruin" (Volz and Rebouças 2016: 104).

The building methods of the Upper Xingu peoples and their worldviews

Before exploring the indigenous architecture of the Upper Xingu, it is relevant to introduce the Xingu Indigenous Park (PIX), created in 1961 to demarcate the protected area for the culture of indigenous Brazilian peoples. The PIX covers and area of 2.8 million hectares and is located in the state of Mato Grosso, in the southern region of the Brazilian Amazon. The "local landscape displays great biodiversity, in an ecological transition region, from the drier savannah and semi-deciduous forests in the south to the ombrophilous forest in the north, includ-ing *cerrados*, fields, floodplain forests, dryland forests and Archaeological Dark Earth forests" (ISA 2021). In this region, the "climate [is] hot, with only two sea-sons: winter, the season of rain and floods, which runs from October to April, with periods of suffocating heat; and summer, the dry season, from May to September, with hot days and cold nights" (Sá and Corrêa 1979: 130).

It is a multi-ethnic region, home to 16 indigenous groups of different ethnici-ties, distributed among the three regions of the territory: the north, known as the *Baixo Xingu* (Lower Xingu), is inhabited by the Kisêdje (Suiá), the Tapayuna, and the Yudjá (Juruna); the central region, called the *Médio Xingu* (Middle Xingu), is home to the Trumai, the Ikpeng (Txikão), and the Kawaiweté (Kaiabi); and in the south, the *Alto Xingu* (Upper Xingu), are peoples who are culturally very similar but speak different languages; these communities show "great similarity in their way of life and work view, especially as they have been connected in a net-work of exchanges, marriages and rituals for centuries" (Almanaque 2011: 21). These peoples are: the Aweti, Kalapalo, Kamayurá, Kuikuro, Matipu, Mehinako, Nafukuá, Naruvôtu, Waurá, and Yawalapiti (ISA 2021).

With such specific regional characteristics, the ten peoples of the Upper Xingu share much in common in terms of the organisation of their villages and the construction of their housing. Architect José Afonso Portocarrero's doctorate research,[4] published in a book, surveyed the types of constructions of some ethnic groups of the Xingu and of peoples from other regions of Mato Grosso state, and will be used as a reference in this chapter to explore the building method used by the indigenous communities of the Upper Xingu.

With that in mind, we return to the artwork *Double Standard*, in which Laís Myrrha stacks organic materials in reference to indigenous housing – a constructive logic that interrelates with nature. Having a harmonious relationship with the environment is not a characteristic exclusive to the indigenous constructions of the Upper Xingu communities, but indigenous peoples in general know how to extract the materials in their surroundings to build their houses without harming nature. Thus, native constructions, regardless of the community, are those developed from the use of local natural materials, taken from nearby native woodland.

From this perspective, studying the native constructions of the Mbyá Guarani people, in the state of Rio Grande do Sul, Zanin (2006)[5] classified them in relation to a full sustainability. For this analysis, I consider these sustainability characteristics as applicable to the buildings of the Upper Xingu indigenous communities, in the sense of full sustainability proposed by the author, involving not only sustainability of the environment, but also the social, cultural, and spiritual aspects, which are not separated in indigenous societies. The author, therefore, lists six characteristics of sustainability: environmental, economic, cultural, social, political, and spiritual.[6]

In our Western society, we need to define the natural relationship that indigenous societies have with nature as sustainable. Since for them, there is no separation between reason and nature, integration between the various spheres is part of their worldviews. Just as the way of taking shelter from the weather is also integrated into all aspects of life, the indigenous house is much more than a place of dwelling. On housing in the Upper Xingu, Portocarrero says,

> For the indigenous peoples the house plays a key role as a place, for they have no other type of construction. The house place is, therefore, a fundamental receptacle of their memories, their worldview, and their history. The houses in the villages are always the same, except for the men's house, which can vary in size, and they take on a character that gives them symbolic value.
>
> (Portocarrero 2015: 200)

Consequently, the village as a whole can be understood both as a mythical and geographic space. The socio-spatial relations with nature develop from the village as a reference. Physically, the village is a group of houses that have the specific characteristics of their residents; however, they share a single construction method. Moreover, Portocarrero, in a mention to Novaes (1983),[7] refers to the architectural aspect of the houses and their spatial organisation in the village, which demonstrate "an idea of a clearly egalitarian society" (Portocarrero 2018: 38).

The village is the set of several *ocas* that have a large central plaza for events and social festivities. Also found in the central plaza are the cemetery and the men's house, or the flute house (which women are forbidden from entering). The houses are built in a circular layout, with a space of 5 to 20 m between them. In the Xinguan house model, the *ocas* have only two doorways to allow light in, one towards the central plaza, the community's social area, and the other to the back into a private area for the residents of the house, where they cook and, in some cases, farm land.

The *oca* is built from materials found locally. Precisely that which will be used is extracted from nature as it is used, avoiding any stockpiling. Everything that is taken from the forest will be used; there is no wastage. The first logs are cut down and used to build the frame, the main columns which support the other structures. Then the other wood is extracted to make the fittings and fastenings so as to give the structure balance through tension points, to withstand the rain and wind, and so on. The materials are extracted during the construction process. Portocarrero reports that the materials used are basically wood for the structure and palm leaves for the roof and sealing.

> The wood is used according to its characteristics: the hardest types are used as pillars and the lightest as beams or secondary structure, generally not in contact with the ground. The straw used for the roof and in the seals is always very similar, but varies depending on the location of the communities; it may be derived from the moriche, babassu, or acuri[8] palm trees, among others. Binding, when still used, is generally from the embira tree.
>
> (Portocarrero 2018: 203)

The inside of the dwelling is frequently swept to keep it clean. Over time, the compacted earth floor merges with the ashes of the fires to form an extremely hard surface, similar to a subflooring.

"The house is always formed by a mesh of thin canes that may or may not be curved, supported on the main frame, composed of raw wooden posts and beams" (Adrião, Pacheco, and Waurá 2020: 94). The houses are oval-shaped, with an area of approximately 13 by 27 m, and there is no distinction between walls and roof. The internal height is about seven to eight metres, with appropriate hot air vents in the upper part, which ensure the houses remain cool during the day, and at night the heat of the fire is conserved to maintain an agreeable internal temperature. The thatched roof provides excellent thermal insulation, and the lack of windows prevents excess sunlight entering the house, keeping it cooler (Portocarrero 2018: 202).

The *oca* is built according to the proportions of the human body. For example, the height of the doorway is the height of the builder, and the width of the doorway measures the same as the distance between the builder's feet with his legs stretched open. Therefore, the house is made in reference to he who will inhabit it. Other important measurements are related to the builder's body measurements: a palm, a foot, the builder's angle of vision from the middle of the chest to the fingertips. Furthermore, the house is understood as anthropomorphic (Portocarrero 2018; Almeida and Yamashita 2013; Troncarelli 2019), where each building element is associated with a part of the human body, like a living being. For example, the central posts are the main supporting frame of the house, and are called "legs". The front of the *oca* to the village plaza is called the "chest", and the rear is called the "back". The "feet" of the *oca* are precisely the row of timbers on the ground, upon which the walls are built. The oval side areas of the *oca*, the private areas of the house, are called the "buttocks". The slats onto which the roof is bound are the "ribs", and the thatch shelter is called "hair". This idea of representing a living being allows the house to be seen as an organic structure. The *oca* breathes thanks to the straw covering, which somehow

> is permeated by the air through infinite micro-interstices created when it is set down, unnoticeable to the naked eye, resulting in a pleasant feeling of thermal comfort generated by the constant and gentle air flow, as if the house were breathing naturally.
>
> (Portocarrero 2018: 202)

Indigenous houses are not only dwellings, they are spaces for social use, where community work is performed and other types of relations established. Three or four generations of the same family might live together in the same space. At the centre of the house, from where one can access the doorways to the plaza and to the back, a fire is constantly alight for cooking food. This fire keeps the house warm at night, and

> the smoke from the night fire inside the house plays a positive role in the maintenance [of the straw roofing], increasing the interval between replace-ment, for it creates a dark inner film coating on the straw, that acts like inverse waterproofing, preventing residues from breaking loose and falling on the hammocks and utensils, and facilitating cleaning.
>
> (Portocarrero 2018: 202)

Also in this central area of the house, provisions (cassava, flour and other food) are kept away from the floor on wooden slats that are suspended on the pillars. At the tips of the house are the private areas, where the residents' hammocks are tied up (ISA 2021; Manual 2019; Adrião, Pacheco, and Waurá 2020: 95). In some houses, straw screens are used as internal walls to divide the spaces and offer the residents greater privacy. However, in general, the interior is ample and open plan, offering space for the various purposes needed for everyday activities.

Portocarrero (2018) reports that this type of house does not require much care and maintenance, only sections of the straw are replaced when deteriorated, and the floor is frequently swept. The straw shelter lasts around four to eight years, after which time it is usually replaced entirely, however the wooden frame is maintained. These houses can last roughly 25 years. To destroy them they are burned down, and new ones can be built on the same site. They tend to be burned down sooner in the event of a cultural factor, such as the death of a resident, or in the event of an insect infestation (cockroaches, for example) between the leaves of the roofing, which must be burned down to ensure sterilisation.

In view of the foregoing description of the building method employed by the Upper Xingu peoples, in *Double Standard*, Laís Myrrha used the construction logic of those peoples, but not the specific materials they use. So, the tower of natural materials was built from stacking materials such as blue gum trunks, bamboo, and sapê thatching. Such materials are easy to find in the urban centres, in this case, the city of São Paulo. The artist opted not to bring native species such as pindaíba, embira, and moriche leaves, for that would mean transporting them more than a thousand kilometres, which would be contrary to the logic of native constructions. Therefore, the artist presents us with the building method based on the underlying logic of indigenous construction. Thus it is possible to understand the indigenous peoples mindset that resembles the circular economy principles. That is, sustainability is not an external demand, in fact, it is an indigenous way of life and interaction with nature.

Civil construction industry

The civil construction industry causes major environmental impact, from the production of the raw materials used to the actual construction process, the sector consumes massive amounts of energy and water resources. To start with, to build a house, usually all the vegetation on the land must be cleared, and earth levelling is often required so the land is fit for the project.

It should also be highlighted that civil construction uses processed materials, that is, materials that are industrialised. Very few materials are extracted and used directly in the civil construction without the need for processing. The production of these materials consumes natural resources and raw materials, also impacting on the environment. Cement, for instance, is an essential ingredient for civil construction, especially due to its use in reinforced concrete, and its production is responsible for major carbon emissions, exacerbating global warming. Huge amounts of energy are consumed to produce glass panes. And the extraction of bauxite for aluminium production also entails the use of high levels of energy and water. The extraction of iron, meanwhile, degrades large areas of vegetation; practically whole mountains are destroyed for iron extraction, and lots of energy is required for transforming it into pig iron. There are also burnt bricks, commonly used in buildings here in Brazil for making walls, and clay tiles, which undergo the same burning process as the bricks, to mention but a few examples.

These examples serve to demonstrate that to build a house one cannot consider only the materials already processed, as if their production that preceded the construction, generated no impact. The supply chain for civil construction begins with extraction, mining, deforestation, the removal and use of water from rivers, and runs through all the pollution-generating industrial production, before the materials are available for the building itself to begin. During the house building, materials are wasted, such as the moulds that are used to make the concrete; and the water is contaminated by toxic substances when equipment and materials are washed. And if the house needs to be demolished, disposing of the large volume of material, of rubble, is a major problem.

Finally, civil construction expends vast amounts of energy and manpower; it demands time, not only for the building itself, but for the manufacture of the materials. Furthermore, it requires specialist labour, and there is not necessarily any direct relationship between those who build the house and those who will live in it. Depending on the building design, heavy machinery is required to execute the job. The construction process, therefore, is costly: the materials are transported from distant places; the natural resources are consumed for the production of the materials; there are elements that are discarded during the process, and in the end none of the material as it has been processed, is incorporated into the environment.

In this construction process, nature is seen as a supplier of raw materials to be exploited: wood, iron ore, bauxite, water, clay, etc. Furthermore, where the house is to be built, nature is seen as an obstacle to be suppressed, even when there are watercourses, springs, etc.

Main environmental impacts of construction:

- Construction consumes 50% to 75% of the natural resources of the world, considering their whole life cycle.
- An estimated 40% of the world's energy is consumed by buildings.
- Approximately 15% of the water resources are consumed by buildings.
- The sector is responsible for 30% to 40% of all CO_2 emissions.
- It is the industry that generates the most waste on the planet. (Sustentarqui 2019).

In the face of environmental impacts, the civil construction sector is timidly in the pursuit of possibilities to include circular economy in its process. Nevertheless, a change in this sector's mindset is more than urgent; even more, it is necessary that circular economy be included in the planning phase of the process.

Finally, despite the severe environmental impact, everything related to the civil construction industry is valued in economic and qualitative terms and is valued as a long-lasting element. However, as the fashion or style changes, renovations are executed and the rubble is discarded. There is very little recycling of the materials, making the disposal of all the rubble and waste material a problem. Whereas civil construction is highly valued, especially in relation to the modern look of the glass and aluminium used, constructions made from organic materials, such as raw clay and straw, are devalued and considered to be of inferior quality. Therefore, they are referred to as housing for less fortunate people, establishing social distinctions.

Considerations

Finally, comparing the two construction processes, Laís Myrrha condemns the impact generated by the take-make-use-dispose-pollute system of the civil construction industry, as opposed to the system integrated into nature used by the indigenous peoples. Moreover, the artist condemns the injustice to which the indigenous system is subjected when it is devalued, due value given by modernity to the perspective of progress and development, without taking into consideration its impacts.

It is in this regard that Mignolo (2017) presents coloniality as the dark side of modernity. Progress brings with it all the devastation of nature, however there is an interest in disassociating this consequence from its image. Therefore, the idea of progress is naturalised, as if it were intrinsic to societies to constantly strive for progress. However, the consequences of progress, such as, for example, social inequality, environmental degradation, and unemployment have been largely disregarded in association to the idea of progress. On the contrary, the thinking that has been disseminated has been that progress would be achieved through development and, paradoxically, all these resulting problems would then be remedied. With this purpose, the drive for development justifies its effects. I refer back to Boaventura de Sousas Santos' term, epistemicide, to address the devaluation of any knowledge that is not aligned to the idea of "progress" and "development". According to Acosta (2016),

> We even deny our historical and cultural roots to modernize and imitate advanced countries. Hence, we deny the possibility of our own modernization. […] For the poor to rise out of their poverty, the rich established that, to be like him, the poor must now pay to imitate him: to buy even his knowledge, marginalizing their own knowledge and ancient practices.
>
> (Acosta 2016: 59)

This is not a question of denying developments, or the importance of technologies, but rather one of reinforcing a side that is kept in the dark, that is, the use of technology for progress as a neocolonial and imperial tool. It seems impossible to think of a world without oil, mining, or agribusiness (Acosta 2016: 239) because a lack of such things would not allow for the so-desired and constantly increasing economic growth. However, it has been shown in the Anthropocene that infinite economic growth is reaching its limits, for it depends on finite resources. Or rather, it has been shown that a limit is being reached here on Earth, which explains the vast investments being directed at the colonisation of Mars. But such exploration carries the same legacy of coloniality; conquering Mars will not signify another way of seeing the world, but rather continuing the same process that began in the sixteenth century, and with the same purposes, continuing to push the problem into the future.

In *Double Standard*, the artist shows a single world and two different approaches of how to deal with it. And the work itself provides possibilities for looking at the world in a different way. Perhaps beginning with the economic logic, as proposed by Acosta (2016), in which the objective is not the accumulation of capital to force constant consumption. To achieve this change, it is important that this new economic proposal be founded on community self-dependence, in order to overcome consumerism and productivism. The economy and society must together change their modes of operation. In this sense, the social and solidarity economy has a lot to contribute. And it would seem that we could learn a great deal from the indigenous peoples and their egalitarian societies. Furthermore, thinking about sufficiency,

> today we are averse to thoughts of slowing, regression, recession, limitation, braking, shrinking, going downward – of *sufficiency*. Anything that recalls

any of these movements in search of an intensive world sufficiency [...] is promptly accused of representing naïve localism, primitivism, irrationality, ill conscience, a feeling of guilt, or even, quite frankly, fascist tendencies.

(Danowski and Viveiros de Castro 2014: 157)

However, sufficiency, as demonstrated by indigenous construction, does not generate accumulation or waste. Everything is taken as and when it is needed for maintaining comfort and keeping nature in balance at the same time.

The intention behind discussing this artwork was to present the possibility of other worldviews, rather than discuss processes. Therefore, citing Danowski and Viveiros de Castro: "Masters of the techno-primitivist bricolage and the political-metaphysical metamorphosis, they [Amerindian collectives] are, actually, one of the possible chances of the future's subsistence" (Danowski and Viveiros de Castro 2014: 159). The aim was to reflect on how much the worldview impacts on everyday processes and offer alternative ways of doing and producing.

Finally, *Double Standard* synthesises how different processes leave their footprints on the environment, just as the worldviews establish the ways of relating to it.

Acknowledgements

This study was financed in part by the Coordenação de Aperfeiçoamento de Pessoal de Nível Superior – Brasil (CAPES) – Finance Code 001.

Notes

1 See Clastres, Pierre, *La société contre l'Estat: recherches d'anthropologie politique.* Paris, Éditions de Minuit, 1974. Trans. Theo Santigo, São Paulo, Cosac Naify, 2003; and *Archéologie de la violence.* (Arqueologia da violência: estudos de antropologia política.) São Paulo: Cosac Naify, 2004. See also Kopenawa, Davi and Albert, Bruce, *A Queda do Céu – Palavras de um Xamã Yanomami.* Trans. Beatriz Perrone-Moisés. São Paulo: Companhia das Letras, 2015.
2 Deuteronomy 25:13–16.
3 Boaventura de Sousa Santos cited Meneses 2007. See MENESES, Maria Paula, "Os espaços criados pelas palavras - racismos, etnicidades e o encontro colonial," in: GOMES, Nilma (org.), *Formação de professores e questão racial* (Belo Horizonte: Autêntica, 2007), pp. 55–75.
4 As a result of his research, Portocarrero published the book *Tecnologia indígena em Mato Grosso: habitação*, which will be used extensively for the descriptions of the indigenous Xingu people's building methods.
5 Nauíra Zanin wrote her masters dissertation at the Engineering School of the Federal University of Rio Grande do Sul (UFRGS) under the title "Abrigo na natureza: construção Mbyá-Guarani, sustentabilidade e intervenções externas" (Shelter in nature: Mbyá-Guarani construction, sustainability and external interventions), in 2006, in which she studied the traditional constructions of the Mbyá-Guarani ethnic group, of Rio Grande do Sul, analysing the maintenance difficulties in the traditional method due to the scarcity of the traditionally used natural elements. The study covered cultural matters of the Mbyá-Guarani, considering its worldview, the *nhande rekó*.

6 See Zanin 2006: 131–132.
7 Portocarrero (2018) makes reference to Novaes, S.C. (Org.) *Habitações indígenas*. São Paulo, Nobel: Edusp, 1983.
8 Acuri or bacuri are the popular name of the palm tree *Scheelea phalerata* (Mart. Ex Spreng.) Burret.

References

Acosta, Alberto. *O Bem Viver: uma oportunidade para imaginar outros mundos. [Good Living: An Opportunity to Imagine Other Worlds]*. Translated into Portuguese by Tadeu Breda. Sao Paulo: Autonomia Literária, Elefante Ed., 2016.

Adrião, J. M. A., Pacheco, T. P. A. P., Waurá, T. A casa Xinguana. *Zeiki, Barra do Bugres*, vol. 1, no. 1, pp. 93–101 (2020).

Almanaque Socioambiental Parque Indígena Do Xingu. *50 anos / Instituto Socioambiental (ISA)*. São Paulo: Instituto Socioambiental, 2011. Available at https://www.socioambiental.org/pt-br/o-isa/publicacoes/almanaque-socioambiental-parque-indigena-do-xingu-50-anos-0. Accessed on 23 April 2021.

Almeida, F. W. Yamashita, A. C. Arquitetura indígena. *Revista de Ciências Exatas e da Terra Unigran*, vol. 2, no. 2 (2013). Available at: https://silo.tips/download/arquitetura-indigena . Accessed on 23 April 2021.

Danowski, Déborah, Viveiros de Castro, Eduardo. *Há mundo por vir? Ensaio sobre os medos e os fins. [The Ends of the World]*. Florianópolis/SC: Cultura e Barbárie: Instituto Socioambiental Ed., 2014.

Deutoronômio. *Conferência Nacional dos Bispos do Brasil, Bíblia Sagrada, Trad. Da CNBB*. São Paulo: Edições Loyola e ed. Paulus, 2001. p. 237.

Instituto Socioambiental/ISA. Povos indígenas no Brasil–Xingu. 2021. Available at https://pib.socioambiental.org/pt/Povo:Xingu. Accessed on 18 April 2021.

Manual De Arquitetura Kamayurá. *Escola da Cidade e Povo Kamayurá. Aldeia Ipawu*. São Paulo: Escola da Cidade, 2019. Available at https://issuu.com/annajubs/docs/190812_casakamayurasingles . Accessed on 23 April 2021.

Mignolo, Walter D. Colonialidade: o lado mais escuro da modernidade. *Rev. bras. Ci. Soc.* [online]. vol. 32, no. 94 (2017). Available at: <http://www.scielo.br/scielo.php?script=sci_arttext&pid=S0102-69092017000200507&lng=en&nrm=iso>. Epub June 22, 2017. ISSN 1806-9053. https://doi.org/10.17666/329402/2017.

Novaes, Sylvia Caiuby. (Org.) *Habitações indígenas*. São Paulo: Nobel/Edusp, 1983.

Portocarrero, José Afonso Botura. *Tecnologia indígena em Mato Grosso: habitação*. Cuiabá/MT: Entrelinhas, 2018.

Quijano, Aníbal. "Bien vivir": entre el "desarrollo" y la des/colonialidad del poder. In: Quijano, Aníbal (org.) *Des/colonialidad y bien vivir – un nuevo debate en América Latina*. Lima: Universidad Ricardo - Editorial Universitaria, 2014. pp.19–33.

Sá, Cristina C. Da Costa e, Corrêa, Eduardo Henrique Bacellar. Habitação indígena no Alto Xingu. In: *Revista Encontros com a Civilização Brasileira, Cinco enfoques sobre a situação Indígena*, n. 12. Rio de Janeiro: Ed. Civilização Brasileira, 1979. pp. 129–142.

Santos, Boaventura de Sousa, Meneses, Maria Paula (Orgs.). *Epistemologias do Sul*. São Paulo: Editora Cortez, 2010.

Sustentaraqui. Impactos ambientais da construção civil. *Notícias*. 05 June 2019. Available at https://sustentarqui.com.br/impactos-ambientais-da-construcao-civil/. Accessed on 23 April 2021.

Troncarelli, Ruth Cuiá. Arquitetura indígena xinguana: um estudo das representações. In: Artur Rozestraten, Gil Barros, Vladimir Bartalini, Karina O. Leitão, Sara Goldchmit (Orgs.) Atas do 3o Colóquio Internacional ICHT, 16 a 18 de abril, 2019, São Paulo, SP, Brasil. Imaginário: construir e habitar a Terra; deformações, deslocamentos e devaneios. São Paulo: FAU/USP, 2019. pp. 705–724. Available at https://sites.usp.br/icht2019/wp -content/uploads/sites/416/2019/07/ARQUITETURA-INDÍGENA-XINGUANA-UM-ESTUDO-DAS-REPRESENTAÇÕES.pdf. Accessed on 23 April 2021.

Volz, Jochen, Rebouças, Júlia (orgs.). *32ª Bienal de São Paulo, Incerteza Viva – Guia*. São Paulo: Fundação Bienal de São Paulo, 2016.

Zanin, Nauíra Zanardo. Abrigo na natureza: construção Mbyá-Guarani, sustentabilidade e intervenções externas. Dissertation (masters), Universidade Federal do Rio Grande do Sul. School of Engineering. Postgraduate Program in Civil Engineering, Porto Alegre, BR-RS, 2006.

12 Beyond circularity

Do we need to shrink and share?

David Ness

Introduction

Despite promoting the UN Sustainable Development Goals (SDGs), the circular economy (CE), net-zero carbon and a "just transition", countries of the Global North continue to consume resources at alarming, unsustainable, and inequitable levels (Hickel 2020; World Bank 2020), while those of the South struggle to meet the basic needs of their rapidly expanding populations.

As Prof Walter Stahel and others have recognised, "the key issue at stake is unbalanced resource consumption at a global level, an issue of global ethics" (Stahel 2008: 508; Factor Ten Institute 2000; Welch and Southerton 2019). They argued that the North should dramatically "shrink" its consumption by 90% so countries within the fast-growing South have a rising "share" of global resources to provide shelter and infrastructure to address systematic inequalities. More recently, this message was reinforced by UNEP (2020) in its *Emissions Gap Report*, while Oxfam (2015) also drew attention to carbon inequality. The rebalancing of emissions and resource consumption has been described, within the context of the ecological footprint, as "shrink and share" (Kitzes et al. 2008).

In tackling climate change, most discussion is limited to cutting GHG emissions from the burning of fossil fuels. However, as Satterthwaite (2009: 547) emphasised, "the dominant cause of global warming is the consumption of goods and services"; this neglects the fact that "reducing resource consumption always reduces CO_2 emissions" (Stahel 2008: 508; Boyd 2019). Thus, resource consumption may be a suitable proxy for emissions and environmental damage (Steinmann et al. 2017).

Moreover, the ODI (2015) emphasised that "poverty eradication cannot be maintained without deep cuts from the big GHG emitters", arguing for climate change and poverty reduction goals to be integrated, namely, "zero poverty, zero emissions". Arguably, zero or responsible consumption could be added, as inequitable resource consumption is the root cause of the main global challenges related to emissions and poverty. The ODI (2015) went further, positing that this strategy should form part of developing and implementing the SDGs. In this regard, SDG 12: "Responsible consumption and production" forms one of the key goals, along with SDG 10: "Reduced inequalities", SDG 11: "Sustainable cities

DOI: 10.4324/9781003255246-12

and communities", and SDG 13: "Climate action" (see Ness 2021: 188–189). However, recent research on climate justice and the built environment, including that led by Klinsky and Mavrogianni (2020), appears to overlook the importance of global equity and ethics in patterns of consumption and related GHG emissions. This chapter seeks to address that gap.

Thus, how can global resource consumption be rebalanced? This gives rise to the research question:

"*How can responsibilities for reducing resource consumption be allocated and shared more equitably, with global resource distribution being rebalanced?*"

The research is based upon a critical examination of the research literature, policies, and extant theories, with the aim of putting forward fresh perspectives for further in-depth examination and testing.

In this regard, the International Resource Panel (IRP 2014: 9) canvassed "equal access to and/or attribution of resource consumption *on a per capita basis* as well as means of ensuring that socio-economic development takes into account the available safe operating space". Exploration of this approach forms the basis of this chapter, especially how it may apply to the built environment. Including buildings and infrastructure, this sector is responsible for around 40% of global resource consumption, emissions, and waste, while also being the cause of much loss of biosphere and environmental despoilation. Moreover, it demonstrates extravagance and inequity in a highly visible way (Figure 12.1).

Figure 12.1 Skyscrapers juxtaposed with informal settlement, India. (Credit: Glen T. Taylor 2015.)

The chapter is structured as follows: firstly, in the second section, the challenge to meet global goals such as the Paris Agreement and the SDGs is outlined. In the third section, the inadequacy of current approaches involving the circular economy is highlighted, with dramatic absolute reduction or "shrinking" consumption being required in wealthier societies, accompanied by "sharing" resources with the needy. In the fourth section, social livelihoods are compared with planetary limits, with the notion of a safe corridor for humanity being explained. The fifth section then examines policies and mechanisms to support this transformation, while the sixth section explains how these may be applied to the built environment. The implications and limitations of the forgoing are discussed in the seventh section, while the eighth section includes closing comments.

The challenges to meet global goals

There is a growing global push for "net zero" climate targets, with many organisations viewing this as a positive signal that the world is on track to meet the Paris Accord. However, a group of climate justice organisations claims "this is far from the truth", and demands close scrutiny (Act!onaid 2020). They argue that "net zero emissions" does not mean "zero emissions", lacks a commitment to real action, and hides deep inequity and injustice. Northern countries and elites may continue to burn fossil fuels while assuming vast plantations in the South will "soak up" their emissions in the distant future, while displacing rural communities from their land (Skelton et al. 2018). Rather than cutting emissions, carbon offsets often shift the burden.

As others have emphasised, the world is not only off-track to reach the Paris target of net zero emissions by 2050, but the "gap" is widening as consumption in wealthier segments of society continues to increase (UNEP 2020). Reducing emissions to a meaningful "real zero" target requires restructuring in multiple sectors, while ecosystems are restored and human rights protected (Act!onaid 2020).

Meanwhile, the global protocol has focused on reducing direct operational emissions (known as Scope 1 and 2), while largely overlooking the importance of Scope 3 consumption-based emissions (WRI, C40, and ICLEI 2015). These include those "embodied" in goods and services, which are presently assigned to poorer "producing" cities. Meanwhile, the richer "consuming" cities, which are ultimately responsible for the emissions of their imported goods, have made the misleading claim that they are "climate neutral". C40 Cities (2018) pointed out that these consumption-based emissions exceed Scope 1 and 2 by a staggering 60%, leading them to argue for reform of the global protocol.

Over ten years ago, Satterthwaite (2009: 564) urged that – to attain the much-needed rapid decrease in GHG emissions globally – "there is an obvious need to focus on rapidly changing the consumption patterns of present (and future) consumers with 'above fair share' GHG emissions". Such warnings have clearly been disregarded, as demonstrated by the per capita and absolute consumption emissions by four global income groups for 2015 (UNEP 2020). UNEP revealed that the richer 1% must dramatically and urgently reduce their consumption emissions by a factor of 30, or 97%, by 2030. Meanwhile, the poorest 50% of society may increase their emissions three-fold, to reach a common per capita emissions target by 2030 for 1.5 degrees temperature rise. Ironically, this lower-income group that contributes the fewest emissions is also the "most vulnerable to the effects of climate breakdown" (Circle Economy 2021: 45).

This is not a new message. Schmidt-Bleek (1993: 114) had previously elevated the issue, arguing that: "the economies of the countries in which, or for which, most of the material flows are presently moved would have to dematerialise by an average factor of 10 in order to allow for a reduction in global material flows by fifty percent". Similar to UNEP's emphasis on rebalancing per capita consumption emissions, Schmidt-Bleek (1993) established a goal of a 50% reduction in global material flows set 50 years hence. As shown in Figure 12.2,

> decreasing material input into the industrialised countries (the "rich" 20 per cent of humanity) can be organised in such a way that a temporary increase and subsequent drop for the countries in the "South" becomes possible within the overall material flow reduction scheme.

Turning to the built environment, the GlobalABC (2020) emphasises that this must decarbonise by 60% by 2030, and 80% by 2050. To meet such targets, a continued preoccupation with reducing operating emissions – comprising 28% of global emissions – will not be enough. As Architecture 2030 (2021) and C40 Cities (2019) have pointed out, the focus must now switch to tackling embodied emissions related to consumption; currently these comprise about 11% of overall emissions, but – given increases in operational energy efficiency – are expected to be equivalent to operational emissions by 2030.

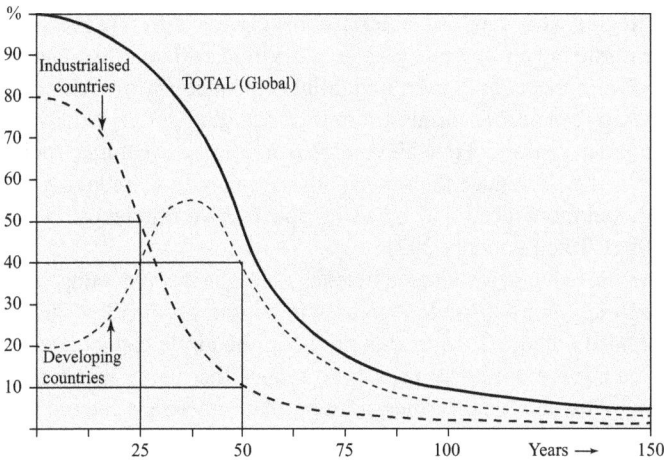

Figure 12.2 Material flow reduction scheme. (Credit: Schmidt-Bleek 1993.)

As with material-based economic growth, more efficient growth in the built environment can lead to a false sense of security, when absolute reductions in energy and material throughput are required. As Rees (2009: 309) highlighted, "new buildings, no matter how green, still add to the total human load". Put simply, if fewer materials are consumed, then embodied carbon will be reduced. However, in tackling embodied carbon, the sector has tended to concentrate on the use of so-called "low carbon materials", whilst avoiding the challenge to build and consume less: "the most environmentally benign building is the one that does not need to be built" (Moffat and Russell 2001).

This points to the need for wealthier societies, already well-endowed with building and infrastructure, to reduce the growth in new stock and new floor area (Ness 2021, 2020). This issue is explored further in the following section.

Inadequacy of current "circular" approaches

The circular economy (CE) is put forward by the Ellen MacArthur Foundation (EMF 2019), Circle Economy (2021), and others as a means of tackling consumption-based emissions. It aims to gain more value from resources by extending their life and keeping them in closed loops, as well as narrowing and slowing the loops. Whilst narrowing flows (use less) and slowing flows (use longer) may be understood in terms of "consuming less" and "sufficiency", many proponents of a CE prefer to focus on reuse and recycling in keeping with their belief in the value of "resource efficiency". This approach, similar to "green growth", is seen as less challenging to "business as usual", when compared with potentially more disruptive calls to consume and use less (Skene 2018; Brown et al. 2014; Stratford 2020).

The *Circularity Gap Report* (Circle Economy 2021: 9) notes that our current economy is only 8.6% circular, leaving "a massive gap", while arguing that "we only need to almost double circularity to close the emissions gap in 2032". Through "smart strategies" and "reduced material consumption", the report argues that global GHG emissions can be cut by 39% and virgin resource use cut by 28%. Similar to the *Emissions Gap Report*, it highlights the inequity of the richest 10% of the globe being responsible for almost half of cumulative CO_2 emissions over the previous quarter-century. To achieve global solidarity, wealthier societies require a major "shift" to reduce their consumption, including "reducing the overall floorspace populations need and by using space more efficiently in the wake of COVID-19" (Circle Economy 2021: 52).

Clearly, strategies involving resource efficiency, circularity, recycling, and re-use are not enough to curtail absolute increases in consumption (EEA 2021). As Allwood (2014: 446) noted, "if demand is growing, the circle cannot remain closed, and it may be a more important priority to reduce the rate at which new material is required". Allwood went further, arguing that rather than having circularity as a goal, "a more pragmatic vision for a material future would be to aim to meet human needs while minimising the environmental impact of doing so". He emphasised that, in parallel with this overall demand reduction, a rebalancing between current rich and poor is required, with rich countries choosing "to want less new material". This leads to the notion of reducing the overall demand for the actual services created by materials, which Allwood termed "the gold medal of environmentally motivated materials management" (Allwood 2014: 450–452).

Therefore, in an effort to "rethink" and "reduce" consumption, we need to switch our focus to services, "so that services, not products, are at the forefront of development" (Carmona et al. 2017: 9); firstly, by questioning service demands, and secondly, comparing alternative pathways for providing the same service with lower resource needs. "This will create space for societal transformation triggered by a more holistic view of sustainable development" (Carmona et al. 2017: 2). Such "rethinking" leads to notions such as a "doughnut economy" (Raworth 2017), where achievement of social livelihoods lies at the core of human development, within overall ecological limits.

Dissatisfied with the prevailing "soft understanding" of a CE, based on closing loops by recycling, the European Environmental Bureau (EEB 2020) emphasised that the objective of a true CE means

> reducing the absolute quantity of natural resources that enter our economy, and reducing the quantity of waste going out...Only with a smaller and slower circle of material throughput will we manage to stay within ecological limits and a safe operating space.

This issue is examined further in the next section.

The "safe operating space"

The concept of a "target corridor" or "safe operating space" for sustainable use and consumption of global resources seeks to balance social needs with ecological limits (Di Guilio and Fuchs 2014). As Bringezu (2015: 25) explained, it considers not only environmental pressures but also social aspects of safe and fair resource use.

O'Neill (2018) quantified the social livelihoods or "basic needs" for over 150 countries, using 11 social indicators or thresholds related to the SDGs, and compared this to what is globally sustainable, using seven environmental indicators related to global limits based on work by the Stockholm Research Centre. O'Neill found that the universal achievement of the social thresholds – including healthy life expectancy, nutrition, sanitation, income, education, and employment – could push humanity past multiple environmental limits. No country achieved all 11 social thresholds without exceeding biophysical boundaries. His findings suggest that some of the SDGs, such as tackling climate change, could be undermined by pursuing those goals focused on growth or high levels of human well-being. In this regard, although wealthy nations may satisfy their citizens' needs, they do so at a level of resource use that far exceeds global environmental limits. This led O'Neill to conclude that radical changes are required, such as removing income inequality. Most importantly, "wealthy nations such as the US and UK must move beyond the pursuit of economic growth, which is no longer improving people's lives…but is pushing humanity ever closer to environmental disaster" (O'Neill 2018: 4).

Hubacek et al. (2017: 361) explained how increasing income leads to increasing carbon footprints, making global targets more difficult to achieve. They concluded that "current carbon-intensive lifestyles and consumption patterns need to enter the climate discourse to a larger extent". In addition, not only did they find huge differences in carbon footprints between countries, but also differences within countries closely linked to differing income levels; even within poor countries, there were "large disparities between the carbon footprints of the rich versus the poor reflecting larger income inequalities in those countries" (Hubacek et al. 2017: 368).

According to Turner and Fisher (2008: 1068), rich countries had already imposed ecological costs of climate change on poor countries that are greater than the foreign debt of the poor countries: "Those bearing these costs include the one billion or so who already lack daily access to safe drinking water, electricity, secure food supplies and basic education". Policies and mechanisms to rebalance these costs are now discussed in the next section.

Policies and mechanisms to shrink and share

The UN Framework Convention on Climate Change (UNFCCC) is based upon the principle of "common but differentiated responsibilities and respective capabilities". The 2015 Paris Agreement requires countries to declare a Nationally Determined Contribution (NDC), with a higher ambition over time. So-called "developed countries" must take the lead by undertaking absolute economy-wide reduction targets, while "developing" countries should continue enhancing their mitigation efforts. The Agreement places greater emphasis on climate-related capacity building for developing countries, while developed countries should demonstrate solidarity via stronger support (UNFCCC 2015a, 2015b, 2019).

As Hickel (2020) has explained, the principle is used to determine differential national responsibilities for mitigation efforts. It can be argued that countries that have contributed more to climate change, via their past industrialisation, are more responsible than those that have contributed less. In this regard, Hickel (2020: 403) argued that "high-income countries must not only reduce emissions to zero more quickly than other countries, but they must also pay down their carbon debts". He presented a "fair-shares" approach for quantifying national debts, rooted in the principle of equal per-capita access to atmospheric commons. This aims to ensure damages sustained by "undershooting" countries due to global warming are paid by "overshooting" countries in proportion to their responsibility.

Kunnas (2011) offered another means of dealing with historical emissions. He suggests splitting the negotiation process into separate blocks. The first would focus on historical emissions, including a "mutual debt cancellation", while the second would deal with present and future emissions, and how to finance adaptation to their consequences. Equating the carbon debts of developed countries with the conventional monetary debts of developing countries could "settle the scores, and start from a clean table". Kunnas acknowledged that a sustainable emissions level has to be divided equally per capita. However, he went further,

arguing that an overdraft could be justified for a transition period, whereby countries emitting over their fair quota would pay for their excess emissions: "By selling emissions quotas to countries needing additional quotas, countries emitting greenhouse gases below their quota…could raise money to be used to finance vital investments in human capital and infrastructure".

Ekins, Meyer, and Schmidt-Bleek (2009: 7) had previously argued that climate policy needs to be complemented with a broader policy focus on resource use: "emissions will fall as policies reduce extractions, but there is no guarantee that reducing emissions will reduce extractions". Similar to the carbon trading mechanism, an international system of marketable permits for use of natural resources was proposed, with targets for reducing resource consumption measured in tonnes per capita. Recognising that establishing complex institutions could pose problems, the establishment of worldwide tax rates on the extraction and import of resources was put forward as an alternative.

Clearly, a mechanism is required whereby emissions credits and resources need to be allocated more equitably or "rebalanced". Rather than establishing a dual system involving trade or taxes in carbon and resource use, with its concomitant complexity, the strengthening and extension of carbon trading and carbon budgets – with a renewed focus on consumption carbon – may be a more realistic way forward. Turning our focus to the built environment, this possibility is examined further in the following section.

Application to the built environment

There is widespread recognition that emissions permits should be allocated on a per-capita basis, with a similar approach having been suggested for material consumption per capita (Bruckner et al. 2012). It has been argued that current material consumption of around 20 tonnes per capita per annum will need to fall to around five to six tonnes by 2050, more than halving resource use in absolute terms. A formidable challenge for which policy has barely begun to be formulated (Ekins, Meyer, and Schmidt-Bleek 2009).

GHG accounting is commonly referred to in terms of "carbon footprint", a widely used metric of climate change impacts. Various authors have examined whether this could be used as a proxy for other environmental impact indicators (Heinonen et al. 2020), while resource footprints may also be suitable proxies for environmental damage and carbon footprints (Steinmann et al. 2017). Laurent et al. (2012) found that buildings and infrastructure displayed strong correlations between carbon footprints and other environmental categories. Thus, we shall focus on the more common metric of carbon footprint.

Considerable research is being undertaken at present to develop carbon benchmarks and budgets for buildings. Much of this work uses a top-down approach, where carbon allocations for countries are "cascaded down" to building levels, resulting in carbon dioxide equivalents (CO_2e) per square metre as a metric. As this fails to take account of the actual size of buildings, it needs to be accompanied by "sufficiency strategies" to reduce floor area per capita (Habert et al. 2020). Hence, carbon per capita or similar metrics are more suitable, such as in determining "sufficient home sizes" in high-income countries (Cohen 2020b).

As the notion of carbon budgets for countries, cities, sectors, and even for buildings becomes more widespread, this is likely to constrain consumption in wealthier societies and constrain new building activity, especially when a carbon price is applied. In this regard, HM Treasury (2013: 11) urged the built environment sector and its clients to "tackle carbon early", as most carbon can be saved during the inception and planning phase of projects. For example, 100% carbon may be saved by "build nothing", that is, by challenging the root cause of the need, and exploring alternative options to achieve the desired outcome, or 80% carbon may be saved by "build less", namely, maximising the use of existing assets, and reducing the extent of new construction required (Figure 12.3). At present, most decarbonisation efforts within the industry are focussed on "build clever", involving the use of low carbon materials, and "build efficiently", although these are only capable of saving 20–50% carbon.

To demonstrate global solidarity in achieving the drastic cuts in carbon required by 2030 and 2050, wealthier societies such as Annexe 1 and OECD members will need to seriously consider non-build options, while adapting and managing the existing stock of buildings and infrastructure (Lanau et al. 2019). In this regard, the response to COVID-19 has shown the way, where services may be provided by digital means and changed patterns of working and living – such as working from home, online education, online retail, and the like, which reduce the demand

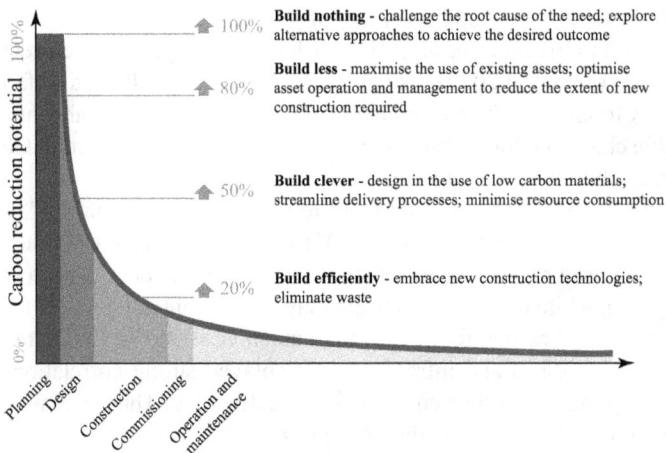

Figure 12.3 Carbon reduction curve – showing the need to tackle carbon early. (Credit: HM Treasury 2013: 11 and Green Construction Board.)

for real estate (Cohen 2020a). Solutions that involve less additional floor area and increased retrofitting will be required (Ness 2020).

This is simply demonstrated by British Land, developer of the Leadenhall Building, London, which recently introduced an internal levy on its carbon emissions. In essence, this involves taxing the carbon on its own projects to create a fund used to retrofit its existing stock. Furthermore, employee bonuses are linked to embodied carbon figures, which has resulted in a preference for retrofit over demolition (Wainwright 2021).

Discussion

Mechanisms based upon carbon and resource allocations, budgets, and exchanges have been presented whereby wealthier societies may "shrink" their consumption – within the built environment and other sectors – and "share" resources, wealth, and expertise with the poor.

The notion of "shrinking" consumption has much in common with the concept of degrowth (Hickel 2019), as well as the principles of "narrowing" and "slowing" material flows that form essential elements (often overlooked) of the circular economy. As Schröder et al. (2019: 9) pointed out, "circular economic principles can contribute to an anchoring of degrowth commitments in an inescapably resource-dependent world".

In this regard, Schröder (2020: 2) has acknowledged that many low- and middle-income countries that rely heavily on "linear" sectors such as extractive industries, manufacture of consumer goods, and agriculture, while exporting these commodities to higher-income countries, may be negatively affected by a shift to circularity. Because this may affect their ability to attain the SDGs, considerations of social justice, solidarity, and inclusivity are important for a circular economy transition in low- and middle-income countries. Comparing this with a low-carbon transition, Schröder places considerable reliance upon "international support" via mechanisms such as compensation, the Solidarity and Just Transition Silesia Declaration, and the World Trade Organization's "Aid for Trade" initiative. However, this may place such countries at the mercy of such mechanisms; is some more fundamental change required?

Bruckner et al. (2012) highlighted the importance of trade in distributing material resources from extraction to consumption countries, thereby widening the gap between low- and high-consuming countries and world regions. The phenomenon of "ecologically unequal exchange", also evident in international trade, allows high-income countries not only to appropriate resources but also to generate a monetary surplus (Dorninger et al. 2020). Such inequalities in resource use "may challenge sustainable and equitable development in non-OECD countries". Hence, extraction and export of primary commodities should be organised "in a way that is economically beneficial to the exporting economies and the local populations" (Bruckner et al. 2012: 574). For example, poorer "producing" countries could compensate for less resource extraction, due to greater circularity in wealthier societies, by adopting extended producer responsibility – thereby adding considerably more value to reduced material exports. Furthermore, as Behrens et al. (2005: 8) advised, it is necessary to "increase incentives to manufacture products in the South instead of merely exporting raw materials", accompanied by increased diversification of industries.

That is, by retaining ownership of materials and goods, providing them as part of a distributed, devolved service system, and taking them back via circular trade – all supported by know-how, capacity building, technologies, and investment from wealthier trading partners. This could "turn the existing system on its head", by positioning lower-income economies in the driver's seat, enabling them to manage resources over their extended life, and providing a pathway to prosperity and global equity.

In addition, the introduction of consumption-based carbon accounting could change the relationship and responsibilities of "producing" cities, many of which are in low-income countries, and wealthier "consuming" cities (Tukker et al. 2020; C40 Cities 2018). The latter would assume responsibility for the emissions produced by exporting cities under a carbon accounting regime that focused upon Scope 3 emissions (Ness and Xing 2020).

Closing comments

This chapter has revisited the pioneering work of the Factor 10 Institute from the 1990s and the often-overlooked concept of "shrink and share" (Kitzes et al. 2008). Recent warnings by UNEP (2020) and the GlobalABC (2020) have echoed the earlier calls for drastic action by wealthier societies to cut consumption-based carbon and respond to the desperate needs of low-income countries for shelter, infrastructure, and services.

It has been shown that approaches based on improved "circularity" and "resource efficiency" are insufficient when demand continues to increase; absolute consumption must be reduced by a new focus on "rethink" and "reduce". Ironically, responses to COVID-19 in wealthier societies, with digital services reducing the need for real estate, may show the way – although the pandemic has exacerbated global inequality and poverty (Khadem 2021).

In this regard, the high consuming built environment warrants special attention. It has also been shown that approaches based on equal per capita allocations of emissions are likely to constrain excessive consumption among wealthier societies, especially when this forms the basis of carbon budgets. We have seen how such budgets can even be applied to buildings, which may further constrain over-consumption (Ness 2021).

The findings have been based upon a critical examination of the research literature, policies, and extant theories, with fresh perspectives having been put forward. Further empirical and in-depth research is required to develop and test the proposals.

In closing, it should be noted that transformation "will require more than simply redesigning production, consumption or accounting models within carbon agreements and systemic structures…solving this issue necessitates a paradigm shift and lifestyle transformation" (Friant 2020). He added: "This entails moving beyond capitalism and building an entirely new system based on solidarity, care, and harmony with the earth", such as exemplified by "degrowth" and "post-development" thinkers (Kothari et al. 2019).

References

Act!onaid. 2020. *Not Zero: How "Net Zero" Targets Disguise Climate Inaction*, Joint Technical Briefing by Climate Justice Organisations, October. Accessed November 6 2021. https://demandclimatejustice.org/wp-content/uploads/2020/10/NOT_ZERO _How_net_zero_targets_disguise_climate_inaction_FINAL.pdf

Allwood, Julian. 2014. "Squaring the Circular Economy: the Role of Recycling Within a Hierarchy of Material Management Strategies." In *Handbook of Recycling: State of the Art for Practitioners, Analysts, and Scientists*, edited by Ernst Worrell and Markus Reuter, 445–477, Amsterdam, Elsevier. https://doi.org/10.1016/B978-0-12-396459-5 .00030-1

Architecture2030. *The 2030 Challenge for Embodied Carbon*. Accessed November 6 2021. https://architecture2030.org/2030_challenges/embodied/

Behrens, Arno, StefanGiljum, Jan Kovanda, and Samuel Niza. 2005. "The Material Basis of the Global Economy: Implications for Sustainable Resource Use Politics in North and South." In *Proceedings of Environmental Accounting and Sustainable Development Indicators*, Charles University Environmental Centre, Prague, September 26–27.

Boyd, Ian. 2019. "Focussing on Cutting Emissions Alone Won't Halt Ecological Decline, We Must Consume Less," *The Conversation*, September 3.

Bringezu, Stefan. 2015. "Possible Target Corridor for Sustainable Use of Global Material Resources," *Resources* 4 (1): 25–54. https://doi.org/10.3390/resources4010025

Brown, Ed, Jonathan Cloke, Danielle Gent, Paul Johnson, and Chloe Hill. 2014. "Green Growth or Ecological Commodification: Debating the Green Economy in the Global South," *Geografiska Annaler: Series B, Human Geography* 96 (3): 245–259. https://doi .org/10.1111/geob.12049

BrucknerMartin, , Stefan Giljum, Christian Lutz, and Kirsten Svenja Wiebe. 2012. "Materials Embodied in International Trade: Global Material Extraction and Consumption Between 1995 and 2005," *Global Environmental Change* 22 (3): 568– 576. http://dx.doi.org/10.1016/j.gloenvcha.2012.03.011

C40 Cities. 2018. *Consumption-based GHG Emissions of C40 Cities*, C40 Cities Climate Leadership Group.

C40 Cities. 2019. *Buildings and Infrastructure Consumption Emissions: in Focus*, C40 Cities Climate Leadership Group, Arup and University of Leeds.

Carmona, Luis, Kai Whiting, Angeles Carrasco, Tania Sousa, and Tiago Domingos. 2017. "Material Services with Both Eyes Wide Open," *Sustainability* 9 (1508): 1-23. https:// econpapers.repec.org/article/gamjsusta/v_3a9_3ay_3a2017_3ai_3a9_3ap_3a1508- _3ad_3a109670.htm

Circle Economy. 2021. *The Circularity Gap Report 2021*, Platform for Accelerating the Circle Economy (PACE), World Resources Institute, Washington, DC.

Cohen, Maurie. 2020a. "Does the COVID-19 Outbreak Mark the Onset of a Sustainable Consumption Transition?" *Sustainability: Science, Practice and Policy* 16 (1): 1–3. https://doi.org/10.1080/15487733.2020.1740472

Cohen, Maurie. 2020b. New Conceptions of Sufficient Home Size in High-Income Countries: Are We Approaching a Sustainable Consumption Transition? *Housing, Theory and Society*, 38 (2), 173-203. https://doi.org/10.1080/14036096.2020.1722218

Di Giulio, Antonietta, and Doris Fuchs. 2014. "Sustainable Consumption Corridors: Concept, Objections, and Responses." *GAIA: Ecological Perspectives for Science and Society* 23 (S1): 184-192. https://doi.org/10.14512/gaia.23.S1.6

Dorninger, Christian, Alf Hornborg, David Abson, Henrik von Wehrden, Anke Schaffartzik, Stefen Giljum, John-Oliver Engler, Robert Feller, Klaus Hubacek, and Hanspeter Wieland. 2020. "Global Patterns of Ecologically Unequal Exchange: Implications for Sustainability in the 21st Century," *Ecological Economics* 179 (C), September. DOI: 10.1016/j.ecolecon.2020.106824

EEA. 2021. "Growth Without Economic Growth," *Narratives for Change*, Copenhagen, European Environment Agency.

EEB. 2020. *A Circular Economy Within Ecological Limits: Why We Need to Set Targets to Reduce EU Resource Consumption and Waste Generation in the New Circular Economy Action Plan*, Brussels, European Environmental Bureau.

Ekins, Paul, Bernd Meyer, and Friedrich Schmidt-Bleek. 2009. *Reducing Resources Consumption: A Proposal for Global Resource and Environmental Policy*, Working Paper, GWS Discussion Paper no. 2009/5, Econstor. http://hdl.handle.net/10419/94417

EMF. 2019. *Completing the Picture: How the Circular Economy Tackles Climate Change*, Cowes, Ellen MacArthur Foundation. Accessed November 6 2021 https://ellenmacarthurfoundation.org/completing-the-picture

Factor 10 Institute. 2000. *Factor 10 Manifesto*, January. http://www.factor10-institute.org/files/F10_Manifesto_e.pdf

Friant, Martin Calisto. 2020. "The Circular Economy: Transformative Vision or Oxymoronic Illusion," *Circular Conversations*. December. https://www.circularconversations.com/research-series-young-researchers/the-circular-economy-transformative-vision-or-oxymoronic-illusion

GlobalABC (Global Alliance for Buildings and Construction). 2020. *Global Status Report for Buildings and Construction*, Global Alliance for Buildings and Construction.

Habert, Guillaume, Martin Röck, Karl Steininger, Antonin Lupisek, Harpa Birgisdottir, Harald Design, Chanjief Chandrakumar et al. 2020. "Carbon Budgets for Buildings: Harmonising Temporal, Spatial and Sectoral Dimensions," *Buildings and Cities* 1 (1): 429–452. http://doi.org/10.5334/bc.47

Heinonen, Jukka, Juudit Ottelin, Sanna Ala-Mantila, Thomas Wiedmann, Jack Clarke, and Seppo Junnila. 2020. "Spatial Consumption-based Carbon Footprint Assessments: A Review of Recent Developments in the Field," *Journal of Cleaner Production* 256: 1-14. https://doi.org/10.1016/j.jclepro.2020.120335

Hickel, Jason. 2019. "Degrowth: A Theory of Radical Abundance," *Real-World Economics Review* 87. http://www.paecon.net/PAEReview/issue87/Hickel87.pdf

Hickel, Jason. 2020. "Quantifying National Responsibility for Climate Breakdown: an Equality-Based Attribution Approach for Carbon Dioxide Emissions in Excess of the Planetary Boundary," *Lancet Planet Health* 4 (9), e399-e404. London: Goldsmiths. https://doi.org/10.1016/S2542-5196(20)30196-0

HM Treasury. 2013. *Infrastructure Carbon Review*, HM Treasury, London.

Hubacek, Klaus, Giovanni Baiocchi, Kuishuang Feng, Raul Castillo, Laixiang Sun, and JinJun Xue. 2017. "Global Carbon Inequality," *Energy Ecological Economics* 2 (6): 361–369. DOI 10.1007/s40974-017-0072-9

IRP (International Resource Panel). 2014. *Managing and Conserving the Natural Resource Base for Sustained Economic and Social Development.* February 7.

Khadem, Nassim. 2021. "Oxfam Report Says Rich Getting Richer and Poor Getting Poorer Amid Coronavirus Pandemic," *ABC News*, January 24.

Kitzes, Justin, Mathis Wackernagel, Jonathan Loh, Audrey Peller, Steven Goldfinger, Deborah Cheng, and Kallin Tea. 2008. "Shrink and Share: Humanity's Present and Future Ecological Footprint," *Philosophical Transactions of the Royal Society B* 363: 467–475. https://dx.doi.org/10.1098%2Frstb.2007.2164

Klinsky, Sonja, and Anna Mavrogianni. 2020. "Climate Justice and the Built Environment," *Buildings and Cities* 1 (1): 412–428.

Kothari, Ashish, Ariel Salleh, Arturo Escobar, Federico Demaria, and Alberto Acosta (eds). 2019. *Pluriverse: A Post-Development Dictionary*, New York, Columbia University Press, October, http://cup.columbia.edu/book/pluriverse/9788193732984

Kunnas, Jan. 2011. "How to Proceed after Copenhagen," *Electronic Green Journal* 1 (31), UCLA, March. http://escholarship.org/uc/item/6bf2k0dz

Lanau, Maud, Gang Liu, Ulrich Kral, Dominik Wiedenhofer, Elisabeth Keijzer, Chang Yu, and Christina Ehlert. 2019. "Taking Stock of Built Environment Stock Studies: Progress and Prospects," *Environmental Science and Technology* 53 (15): 8499–8515. https://doi.org/10.1021/acs.est.8b06652

Laurent, Alexis, Stig Olsen, and Michael Hauschild. 2012. "Limitations of Carbon Footprint as Indicator of Environmental Sustainability," *Environmental Science Technology* 46 (7): 4100–4108. https://doi.org/10.1021/es204163f

Moffat, Sebastian, and Peter Russell. 2001. "Assessing the Adaptability of Buildings," *Energy Related Environmental Impact of Buildings*, Annex 31, IEA.

Ness, David. 2020. "Growth in Floor Area: The Blind Spot in Cutting Carbon," *Emerald Open Research*, January. https://doi.org/10.35241/emeraldopenres.13420.3

Ness, David. 2021. *The Impact of Overbuilding on People and the Planet.* 2nd ed., Cambridge Scholars, Nottingham. https://www.cambridgescholars.com/product/978-1-5275-7470-0

Ness, David, and Ke Xing. 2020. "Consumption-based and Embodied Carbon in the Built Environment: Implications for APEC's Low Carbon Model Town Project," *Journal of Green Building* 15 (3): 67–82. https://doi.org/10.3992/jgb.15.3.67

ODI (Overseas Development Institute). 2015. *Zero Poverty, Zero Emissions: Eradicating Extreme Poverty in the Climate Crisis.* London, September.

O'Neil, Dan. 2018. "Is it Possible for Everyone to Live a Good Life Within Our Planet's Limits", *The Conversation*, February 8.

Oxfam. 2015. "Extreme Carbon Inequality: Why the Paris Climate Deal Must Put the Poorest, Lowest Emitting and Most Vulnerable People First", *Media Briefing*, December 2.

Raworth, Kate. 2017. *Doughnut Economics: Seven Ways to Think Like a 21st Century Economist*, London, Random House.

Rees, William. 2009. "The Ecological Crisis and Self-delusion: Implications for the Building Sector," *Building Research and Information* 37 (3): 300–311. http://dx.doi.org/10.1080/09613210902781470

Satterthwaite, David. 2009. "The Implications of Population Growth and Urbanisation for Climate Change," *Environment and Urbanisation*, International Institute for Environment and Development, 21 (2): 545–567. https://doi.org/10.1177%2F0956247809344361

Schmidt-Bleek, Friedrich. 1993. "Factor 10." Chapter 5 in *MIPS: the Fossil Makers*, Carnoules, Factor 10 Institute. http://www.factor10-institute.org/files/the_fossil_makers/FossilMakers_Intro.pdf

Schröder, Patrick. 2020. *Promoting a Just Transition to an Inclusive Circular Economy*, Energy, Environment and Resources Programme, London. Chatham House, April.

Schröder, Patrick, Magnus Bengtsson, Maurie Cohen, Paul Dewick, Joerg Hoffstetter, and Joseph Sarkis. 2019. Degrowth Within: Aligning Circular Economy and Strong Sustainability Narratives, *Resources, Conservation & Recycling* 146: 190–191. 10.1016/j.resconrec.2019.03.038

Skelton, Alasdair et al. 2018. "10 Myths about Net Zero Targets and Carbon Offsetting, Busted," *Climate Change News*, December 11.

Skene, Keith. 2018. "Circles, Spirals, Pyramids and Cubes: Why the Circular Economy Cannot Work," *Sustainability Science* 13(6): 479–492. https://link.springer.com/article/10.1007/s11625-017-0443-3

Stahel, Walter. 2008. "Global Climate Change in the Wider Context of Sustainability," *The Geneva Papers* 33 (3): 507–529. doi:10.1057/gpp.2008.21

Steinmann, Zoran, Aafke Schipper, Mara Hauck, Stefan Giljum, Gregor Wernet, and Mark Huijbregts. 2017."Resource Footprints are Good Proxies of Environmental Damage," *Environmental Science and Technology* 51 (11): 6360–6366. https://doi.org/10.1021/acs.est.7b00698

Stratford, Beth. 2020. "Green Growth vs Degrowth: Are We Missing the Point?" *openDemocracy*, December 4. https://www.opendemocracy.net/en/oureconomy/green-growth-vs-degrowth-are-we-missing-point/

Tukker, Arnold, Hector Pollitt, and Maurits Henkemans. 2020. "Consumption-based Carbon Accounting: Sense and Sensibility," *Climate Policy* 20(sup1): S1–S13. https://doi.org/10.1080/14693062.2020.1728208

Turner, Kerry, and Brendan Fisher. 2008. "To the Rich Man, the Spoils," *Nature* 451 (7182): 1067–1068.

UNEP. 2020. *The Emissions Gap Report*, United Nations Environment Program, Nairobi.

UNFCCC. 2015a. *Key Aspects of the Paris Agreement*, UN Framework Convention on Climate Change, Bonn.

UNFCCC. 2015b. *What is the Paris Agreement?* UN Framework Convention on Climate Change, Bonn.

UNFCCC. 2019. *Report of the Conference of the Parties on the Twenty-fifth Session*, Madrid, UN Framework Convention on Climate Change, December 2–6.

Wainright, Oliver. 2021. "The Dirty Secret of So-called 'Fossil-fuel Free' Buildings," *The Guardian*, April 3.

Welch, Daniel, and Dale Southerton. 2019. "After Paris: Transitions for Sustainable Consumption," *Sustainability: Science, Practice and Policy* 15 (1): 31–44. https://doi.org/10.1080/15487733.2018.1560861

World Bank. 2020. *Poverty and Shared Prosperity 2020: Reversals of Fortune*, World Bank, Washington, DC.

WRI, C40 and ICLEI. 2015. *Greenhouse Gas Protocol: Global Protocol Community-scale Greenhouse Gas Emission Inventories: An Accounting and Reporting Standard for Cities*, World Resources Institute, C40 Cities, and Local Governments for Sustainability, Washington, DC.

Index

For Product Safety Concerns and Information please contact our EU
representative GPSR@taylorandfrancis.com
Taylor & Francis Verlag GmbH, Kaufingerstraße 24, 80331 München, Germany

www.ingramcontent.com/pod-product-compliance
Lightning Source LLC
Chambersburg PA
CBHW060257220326
41598CB00027B/4135